GOAT HUSBANDRY

Ruth Goodwin studied zoology at university and has been keeping her own goat herd since 1961. A life member of the British Goat Society, she has held the position of Scientific Liaison Officer for 25 years, and edited the Society's monthly journal from 1985 to 1990. For the past three years she has written a monthly magazine article on goatkeeping and she belongs to a large number of goat societies, and to the Society of Dairy Technology. In 1992 she was elected a Vice-President of the Surrey Goat Club.

also published by Faber and Faber

Small-scale Poultry Keeping: A Guide to Free-range Poultry Production by Ray Feltwell

Small-scale Sheep Keeping by Jeremy Hunt

The Complete Herbal Handbook for Farm and Stable by Juliette de Baïracli Levy

The Complete Herbal Handbook for the Dog and Cat by Juliette de Baïracli Levy

The Illustrated Herbal Handbook for Everyone by Juliette de Baïracli Levy

Cats – Naturally by Juliette de Baïracli Levy

Goat Husbandry

by DAVID MACKENZIE

Revised and Edited by RUTH GOODWIN

faber and faber
LONDON · BOSTON

First published in 1957
by Faber and Faber Limited
3 Queen Square London WC1N 3AU
Second edition 1967
Third edition 1970
Reprinted 1972, 1974, 1975, 1976 and 1978
Fourth edition 1980
Reprinted 1985, 1990
Fifth edition 1993

Photoset by Wilmaset Ltd, Wirral
Printed in England by Clays Ltd, St Ives plc

A CIP record for this book is available from the British Library

ISBN 0–571–16595–8

10 9 8 7 6 5 4 3 2 1

Contents

Illustrations

Plates

Figures

Tables

Preface to the First Edition

Certain references in this Preface are to material omitted from later editions

Nearly twenty years ago, I retired into a converted hen house with a milking pail, a book of instructions, and an elderly goat of strong character. There was milk, among other things, in the pail when the goat and I emerged at last, with mutual respect planted in both our hearts. The book of instructions was an irrecoverable casualty.

No book is a substitute for practical experience, but books become more important as animals become more productive; the modern farmer and smallholder has to remember so much, so often and so quickly that each class of farm livestock requires to be accompanied by a comprehensive work of reference as a passport to the modern farmyard.

Holmes Pegler's *The Book of the Goat* was published first in 1875, and is at present in its ninth edition (now out of print); *Modern Milk Goats*, by Irmagarde Richards, was published during the first quarter of the present century in the USA. In the meantime there have been published on both sides of the Atlantic several excellent practical handbooks on goatkeeping, goat farming and goat fancying, but no work with so wide a scope as *The Book of the Goat*. The omission is one of several handicaps to an expansion of commercial goatkeeping and always struck me as surprising until I found myself involved in the attempt to fill it.

The fact is that the study of the goat has commanded rather less scientific attention than the study of mice and guinea-pigs. The practical of goatkeeping in Britain, though successful in establishing world milk-production records and a vigorous export trade, is so highly variable, individual and empirical that it is almost indescribable. In search of information from scientific and technical institutions, the writer on goats is liable to be treated with the respect accorded to all things pertaining to the goat in Britain. While the majority of the leading exponents of goatkeeping practice in this country are eager to help, few have the time or temperament to measure and record their own practice with useful accuracy.

At several stages in the writing of this book I should have quailed before the obstacles so presented, but for the encouragement of Lawrence D. Hills and his fertile suggestions.

To Professor S. A. Asdell of Cornell University, USA, and to Professor C. W. Turner of Columbia University, USA, I am particularly grateful, not only for guiding me to the scientific work which laid the basis of the study of goat nutrition, but for their kindness to a stranger.

Many of my readers may agree with me in regarding the section dealing with problems of mineral imbalance in the goat as of greater practical use to the breeder than any other in this book; they will wish to join me in thanking Dr E. C. Owen of the Hannah Dairy Research Institute, and Brynmore Thomas of the Durham University School of Agriculture, who have provided me with much of the material on which this section is founded.

In my chapter on 'Goats' Milk in Human Nutrition', as a stockman and farmer, I found myself on rather unfamiliar ground; I am particularly grateful to Dr J. B. Tracey for reading my chapter and enabling me to enrich it with the fruits of his unique joint experience of goats and medical practice. For guidance in the special literature on this subject, I am indebted to Dr Mavis Gunther, MA, MD, of University College Hospital, London, whose kindness in reading and criticizing the chapter has greatly strengthened it. I was also assisted here by the advice of Sister E. Morrison, of *Nursery World*, based on her incomparable experience of the feeding troubles of 'difficult' infants, and by the classic work and personal assistance of F. Knowles, Honorary Analyst to the British Goat Society.

I would also express my thanks to many busy men of science who found time to give me help on technical points: to Stephen Williams, MSc, farms' manager to Boots Pure Drug Company, for sources of information on the food capacity of the cow; to the Veterinary Advisory Department of Boots Pure Drug Company for information on the relationship between parasitic worm infestations and Vitamin B deficiencies; to James A. Paterson of the Scottish Milk Records Association; to Dr Fraser Darling; to C. A. Cameron, managing director of Alginate Industries (Scotland); to G. Kenneth Whitehead, Withnell Fold; to the Mond Nickel Company for information on cobalt, and to the Iodine Educational Bureau for a good deal of material on breeding troubles.

The short survey of industrial waste products suitable for goat feeding obliged me to trespass on the kindness of the technical executives of a number of firms; I thank Dr V. L. S. Charley of H. W. Carter & Company for a particularly happy correspondence on the use of blackcurrant pomace, and for generous assistance in investigating the matter; R. B. Norman of H. J. Heinz Company, W. E. Rhodes of Chivers & Sons, R. S. Potter of William Evans & Company, and R. E. Harris of Calindus Food Products for their helpful information.

I have particularly enjoyed the correspondence with friends of the goat in other countries, and hope that the list of names and addresses in Appendix I [now App. 2] will not only aid the circulation of knowledge on goatkeeping but enable travelling goatkeepers to make contacts over a considerable proportion of the world. In compiling this list and for information on world goatkeeping I am indebted to Tanio Saito of Uwajima, Japan, S. de Jogn Szn., Secretary of the Goat Society of the Netherlands, Sigurd Andersen, Editor of the *Journal* of the Danish Goat Society, Mr Robert W. Soens, Secretary of the American Milk Goat Record Association, Monsieur C. Thibault, Director of the Station for Research in Animal Physiology, Jouy-en-Josas, France, Mrs du Preez of Cape Province, South Africa and Dr Finci of the Department of Agriculture of Israel.

Of goatkeepers in this country I must first express my gratitude to Miss M. F. Rigg, the ever-helpful Secretary of the British Goat Society, whose assistance has been spread over the several years during which this book was being prepared, and has been indispensable. I acknowledge with gratitude the permission of the Committee of the British Goat Society to make use of photographs and many extracts from articles first published in the British Goat Society year books and monthly journals. I thank the many contributors to these richest sources of goat knowledge.

Of individual goat breeders I owe a special debt to Mrs J. Oldacre, Hanchurch Yews, for the photographs of yarded goats, the plans of a model goat-yarding system and goat garden prepared by her architect husband, and for some of the most carefully recorded information on practical goatkeeping that it has been my pleasure to use; J. R. Egerton for many beautiful photographs and for information on RM5 Malpas Melba; to Lady Redesdale for her cheese recipe; to Mrs Jean Laing, Moorhead, Newton Stewart, for photographs, for information on her feeding methods and for providing a shining example of economic methods of management; to Miss Mostyn Owen, Mrs Margaret Train and Mrs J. D. Laird for photographs; and to Miss Jill Salmon, Cothlan

Barn, for some fine photographs and for that rare and precious material – accurately recorded information on feeding for high yields.

I am grateful to the Ministry of Agriculture and Fisheries and the Department of Agriculture for Scotland for the trouble they have taken to prepare the statistics of goat population in Appendix 2 [now deleted], and acknowledge with thanks the permission of the Director of Publications at Her Majesty's Stationery Office for permission to use material from 'Rations for Farm Livestock, Bulletin 48', and a design from 'Farm Gates' (Fixed Equipment on the Farm, Leaflet 8), as the basis of the goat-proof latch, illustrated in Fig. 10.

The photograph of the Wolseley Electric Fence, and the drawings of the Gascoigne Goat-Milking Machine, are kindly contributed by the firms concerned.

The reproduction of Bewick's Goat is by permission of the Victoria and Albert Museum.

The remaining drawings are by Kenneth Hatts of Bournemouth Art College. I feel more grateful to him each time I look at them.

The main brunt of book writing is borne by the author's household; I think particularly of my small son who, on the day after the manuscript was posted to the publishers, systematically destroyed the remaining stocks of stationery in the house to prevent me writing another.

Glen Mhor, Kishorn, Wester Ross DAVID MACKENZIE
8 April 1956

Preface to the Second Edition

Certain references in this Preface are to material omitted from later editions

To the many goat breeders from many countries, who have sent me news, comments, problems and criticisms, are due my thanks and the credit for most of the improvements to be found in the new edition.

I am particularly grateful to Miss M. F. Rigg, Secretary of the British Goat Society, for helping me in various ways, and for assistance in investigating the circumstances and the extent of the serious decline in the productivity of goats in Britain.

To have had the help of Dr M. H. French, Chief of the Animal Production Branch of the Food and Agriculture Organization of the United Nations, is a privilege which has enabled me to give an up-to-date evaluation of the importance of goats in the agriculture of developing countries, and to shatter the image of the Destructive Mediterranean Goat, which lies at the heart of so much anti-goat prejudice.

The therapeutic use of goats' milk, especially in cases of allergy to cows' milk, accounts for an increasing proportion of goats'-milk production in Britain and the USA. During the past ten years there has been a volcanic eruption of scientific evidence about the extent and gravity of cows'-milk allergy. This alarming material is very relevant to the future of commercial goat breeding and dairying, and I have tried to present it in a form usable by a goat farmer and by respectable medical men. Dr L. Sutherland, MB, ChB, DPH, DTM & H, has done her best to guide my pen in the way of objective truth in this matter. I do not wish to saddle her with any responsibility for statements made in this book, but with my gratitude.

By calling my attention to the newly discovered method of deodorizing male goats and eliminating the occasional 'goaty' flavour from goats' milk, Miss Jill Salmon has increased the load of gratitude I already owe her, and provided the answer to many a goatkeeper's prayer.

I confess that I lack the patience and devotion to listen for years on end to all the troubles, big or little, of every goatkeeper who cares to write or phone, and to search out a helpful and sympathetic answer to each one. The extent of my practical knowledge suffers as a result, but this book suffers relatively little, because I have been able to call on the prodigious volume of such practical knowledge that Mrs Jean Laing has accumulated. Like many goatkeepers in Scotland, I am thankful for it.

Finally a word of thanks to readers who are going to write to tell me what I have done wrong this time, and to ask me questions to which I do not know the answer. They keep the book alive.

Bridge of Urr, Kirkcudbrightshire DAVID MACKENZIE
30 November 1965

Preface to the Fourth Edition

When I agreed, with some reluctance, to attempt the task of updating David Mackenzie's classic text on goatkeeping, which has become known to goatkeepers throughout the world as their 'bible', I had little idea what I was letting myself in for. Without a generous response from those who so kindly made time to help me, the task would have been impossible.

Where additions and amendments have been necessary, I have followed as nearly as possible the beliefs of the author. This has meant that I have in some cases seemed to put words into the author's mouth, as it were, but this has proved to be the most satisfactory way of maintaining the flow of the text without editorial insertions. Where I have omitted material included in previous editions this has, in the main, been on account of recent research and developments, and in order to present goatkeeping in the light of today's conditions.

While appreciating the enormous contribution of the dedicated master breeders, whose skill has given us the high quality, sound conformation and long lactation of the dairy goat in Britain, and on whom we so much depend for our stud goats, David Mackenzie was primarily concerned with the goat as an economic household provider. He was also much concerned with the importance of goats' milk to those suffering from certain allergic conditions. His chapter on the nutritive and curative properties of goats' milk was thoroughly revised by him for the second edition, shortly before his death, and as it seems certain that he would have wished to revise it further for publication today, this chapter has been omitted. I have, therefore, at the publisher's request, added a short Appendix on 'Allergies and Goats' Milk'. I have added too an Appendix containing a few 'Notes for Novices' [now deleted] where advice on some of the problems commonly encountered is gathered together for easy reference.

My thanks are due to the British Goat Society for allowing reproduction of material from their year books and journals, and to Bob Martin, the BGS Shows Secretary, for his Appendix giving details of the revised regulations for Milking Competitions [now deleted]; to those who supplied new photographs; to Philippa Awdry, Pamela Grisedale, Bryce Reynard and Robert Haslam, for accounts of their goat-farming systems; to Mrs Leueen Hill of Redruth for her recipes for smallholder cheese and lactic acid curds; to Malcolm Castle and Paul Watkins for the information in Appendix 4 [now deleted]; to Peter Wray, MRCVS, for his advice on disease and accidents; to Patricia Sawyer for taking the trouble to write out all those tricky names and addresses in various countries, and for her notes on export; and finally to Ann Rusby, who struggled so valiantly on her typewriter at home, and to Barbara Ellis for her painstaking editorial help, both of whom managed somehow to decipher my handwriting.

January 1980 JEAN LAING

Preface to the Fifth Edition

Since *Goat Husbandry* was revised in 1980, the goatkeeping scene in the UK has changed almost beyond recognition. New breeds and an increase in understanding of goat management are welcomed, but not the new diseases imported from overseas. I wonder what David Mackenzie would have made of the legislative strait-jacket, being pulled ever tighter around today's Euro-goatkeeper? To set out an account of all the new regulations in his book seems almost to defile the pages, and yet, since Britain signed the Treaty of Rome and Europe became gripped by legislation mania, these matters have to be studied and complied with. No doubt in time, like all changes, the rules will be accepted without question.

The acceptance of goat farming into mainstream agriculture has enabled me to rewrite Chapter 3, 'The Prospects for Goat Farming', while research results from around the world have made it necessary to replace the two chapters on feeding (6 and 7). Increased scientific and veterinary knowledge have been reflected in alterations to the chapters on breeding and disease, (8, 9 and 10), and much of the new legislation is described in the chapters on milking practice and the export trade (11 and 17).

Cashmere, Cashgora and Angora goats are dealt with principally in the chapter on leather and fleece (14), while harness goats form the subject of a new chapter (18).

The lists of overseas addresses, suppliers of requisites and the bibliography have required extensive updating. The entire text has been subject to alterations here and there, I hope without losing the character of this unique and classic work.

Albury, Surrey RUTH GOODWIN
June 1992

Introduction

(1956, revised 1965)

When man began his farming operations in the dawn of history, the goat was the kingpin of the pastoral life, making possible the conquest of desert and mountain and the occupation of the fertile land that lay beyond. The first of man's domestic animals to colonize the wilderness, the goat is the last to abandon the deserts that man leaves behind him. For ever the friend of the pioneer and the last survivor, the goat was never well loved by arable farmers on fertile land. When agriculture produces crops that man, cow and sheep can consume with more profit, the goat retreats to the mountain tops and the wilderness, rejected and despised – hated too, as the emblem of anarchy.

During the last hundred years much has been done to improve the productivity of goats in Northern Europe, North America and Australia, where the modern dairy goat can convert the best of pastures and fodder crops into milk as efficiently as the modern dairy cow. Like most small production units, the goat is expensive with labour, but in its use of raw materials it rivals the cow, even when the raw materials are those best suited to the cows' requirements.

In developing countries, where there is normally an embarrassing surplus of suitable labour, the high labour requirement of goat dairying is a social virtue. If the land be arid or mountainous, the goat may prove to be the only economic source of milk. The authors of national development plans and international aid projects are not so starry-eyed today as they were ten years ago; the cow dairy farm in the desert and the steel mill in the jungle are giving way to reality and goat improvement schemes.

In the advanced countries, medical research has been discreetly lifting the lid off the consequences of our peculiar practice of snatching the newborn infant from its mother's breast and fostering it on the first available cow. The consequences may be grim, or may persist throughout life, unless goats' milk comes to the rescue. In Britain and the USA there is a growing demand for goats' milk for therapeutic use, which cannot afford to be deterred by the higher labour costs of goats'-milk production.

Unfortunately, this new resurgence of interest in goat dairying may, in the present state of goat breeding, wreak havoc. The last great resurgence of goatkeeping met the wartime challenge of food shortage, and culminated, in 1949, in the legal black market in dairy produce. From that date we mark a steady decline in the quality of the yields of pedigree goats in Britain, a decline which still continues. The new goatkeepers of 1949 were taught to feed their goats as miniature cows, and proceeded to breed them selectively for their response to this feeding. With honourable exceptions, the goat breeders of 1965 maintained this destructive tradition. An expansion of goat breeding within this tradition can only result in establishing a strain of goats which do, in fact, perform like miniature cows. A good cow, miniaturized to goat size, could produce no more than 6 pints a day on the best feeding. A good goat can produce twice as much.

The relationship between size and efficiency in all productive farm animals is so well established in both theory and practice that, when confronted with the performance of the modern dairy goat, the nutritional scientist and the farmer tend to regard it as a somewhat indecent Act of God, unrelated to His regular arrangements. For the rule is adamant: provided the feeding is sufficient, the big animal must outyield the little one; the big one has a smaller surface area in proportion to its bulk and potential food capacity, and so uses less of its food to keep itself warm and more to make meat or milk. Friesians replace Ayrshires as pastures are improved; low-ground sheep are bigger than mountain breeds; every beast, ideally, is as big as its pasture permits. But fifteen 1-cwt [50·8 kg] goats will make rather more milk out of the ration of a 15-cwt [762 kg] Friesian cow than the cow can. Yet the rule is unbroken; for it applies only between animals of the same species: a kind Providence has decreed that goats are very far from being miniature cows.

A goat, however 'modern' and 'dairy-bred', is a goat, a member of the species familiarized in nursery picture books and biblical illustrations, target of laughter and abuse for countless centuries, Crusoe's salvation and mankind's first foster mother, the Common Goat.

The processes of history have greatly reduced the goat in Britain; agricultural textbooks have exiled the hardy ruffians for half a century; scientists have used the modern dairy goat as an expendable model cow, but done little to investigate the basic attributes of the goat as such. The purpose of this book is to drag this half-mythical creature out into the light of present-day animal husbandry, that we may know it, use it and care for it more effectively.

We must begin by evicting from our minds the false analogies between goat and cow and between goat and sheep. We can hardly do better than refer back to Thomas Bewick's *History of Quadrupeds*, published at the beginning of the nineteenth century, when the Common Goat was still common in Britain.

> This lively, playful and capricious creature occupies the next place in the great scale of nature; and though inferior to the sheep in value, in various instances bears a strong affinity to that useful animal.
>
> The Goat is much more hardy than the sheep, and is in every respect more fitted to a life of liberty. It is not easily confined to a flock, but chooses its own pasture, straying wherever its appetite or inclination leads. It chiefly delights in wild and mountainous regions, climbing the loftiest rocks and standing secure on the verge of inaccessible and dangerous precipices; although, as Ray observes, one would hardly expect that their feet were adapted to such perilous achievements; yet, upon nearer inspection we find that Nature has provided them with hoofs well calculated for the purpose of climbing; they are hollow underneath, with sharp edges like the inside of a spoon, which prevent them from sliding off the rocky eminences they frequent.
>
> The Goat is an animal easily sustained, and is therefore chiefly the property of those who inhabit wild and uncultivated regions, where it finds an ample supply of food from the spontaneous production of nature, in situations inaccessible to other quadrupeds. It delights in the healthy mountain or the shrubby rock, rather than the fields cultivated by human industry. Its favourite food is the tops of the

boughs or the tender bark of young trees. It bears a warm climate better than the sheep, and frequently sleeps exposed to the hottest rays of the sun.

The milk of the Goats is sweet, nourishing and medicinal, and is found highly beneficial in consumptive cases; it is not so apt to curdle on the stomach as that of the Cow. From the shrubs and heath on which it feeds, the milk of the Goat acquires a flavour and wildness of taste very different from that of the Sheep or Cow, and is highly pleasing to such as have accustomed themselves to its use; it is made into whey for those whose digestion is too weak to bear it in its primitive state. Several places in the North of England and in the mountainous parts of Scotland are much resorted to for the purpose of drinking the milk of the Goat; and its effects have been often salutary in vitiated and debilitated habits.

In many parts of Ireland and in the Highlands of Scotland, their Goats make the chief possessions of the inhabitants; and in most of the mountainous parts of Europe supply the natives with many of the necessaries of life: they lie upon beds made of their skins, which are soft, clean and wholesome; they live upon their milk and oat bread; they convert part of it into butter and some into cheese. The flesh of the Kid is considered as a great delicacy; and when properly prepared is esteemed by some as little inferior to venison.

The Goat produces generally two young at a time, sometimes three, rarely four; in warmer climates it is more prolific and produces four or five at once. The male is capable of propagating at one year old and the female at seven months; but the fruits of a generation so premature are generally weak and defective; their best time is at the age of two years or eighteen months at least.

The Goat is a short-lived animal, full of ardour, but soon enervated. His appetite for the female is excessive, so that one buck is sufficient for one hundred and fifty females.

Thomas Bewick's account of the goat suffers little from the passage of nearly two hundred years. The wildness of taste which he attributes to goats' milk can be tamed by dairy hygiene, mineral supplement and surgical operation, but many newcomers to goats' milk can still catch his meaning. The resident population of the 'wild and uncultivated regions' has been eroded by hunger and administration, but the 'healthy mountain and shrubby rock' are still good dairy pastures for a goat. Bewick's few paragraphs contain clues to the peculiarities of goat digestion, housing requirements, and control, and to the phenomenal productivity of the modern dairy goat.

In following up these clues in subsequent chapters, the assumption is that goat farming can and will develop into a considerable branch of agriculture. As such, goat farms must be mainly in the hands of established farmers with a general knowledge of crop and stock, and utter ignorance of goats. Such knowledge and ignorance is assumed; but a chapter on cropping for goats is included to help the domestic and small-scale goatkeeper in whose hands the goat tends to be most profitable.

CHAPTER ONE

The Place of the Goat in World Agriculture

Cow, sheep and goat, all provide man with meat and milk. At times we are inclined to think and act as though they were rivals for that dubious honour. In fact they are prehistoric grazing companions, who need each other's help to make the most of available pastures.

The natural covering of the earth ranges down from the unbroken canopy of high forest to the small, ephemeral herbage of the desert, hot or cold. the permanent natural grasslands lie next to the desert fringe: steppe, prairie, pampas, mountain top and sand dune grow only grass because the soil is too thin, dry, cold, to grow anything bigger. Their vast extent supports large herds, but the typical stocking capacity of such pastures is only a cow and calf to thirty acres; under these conditions sheep, cow and goat are as competitive as they are complementary and the productivity of the land is inevitably low.

The best natural pastures are temporary, being stolen from woodland by fire, drought, flood and storm. There is good natural grazing too on savannah-type land, where soil and climate maintain a precarious balance between 'bush' and grasses. Many pastures are man-made; the best of them, in New Zealand, Holland, the English Midlands, being derived from marsh or woodland. Temporary grazings, whether natural or man-made, can be maintained only by the co-ordinated efforts of grazing stock and by man.

The reversion of such temporary pastures to scrub, woodland and forest is pioneered by coarse grasses and unappetizing weeds, rushes, thistles, brackens, nettles, etc; these are followed by still more repellent small shrubs, bramble, briar, gorse and thorn. Within the bridgehead so established, the windblown seed of light-leaved trees can germinate and grow, to provide a refuge for beasts and birds that carry the seed of the forest giants, in whose shade all rivals perish.

The first line of pasture defence is the sheep, whose split lip enables it to bite herbage down to soil level. The sheep catches the toughest of invaders in their seedling stage, while they are still tender and nutritious. Its daily capacity for food intake is smaller in relation to its size than of its companions, so its grazing habits are more selective; the sheep generally avoids coarser vegetation, but exercises some control of established shrubs by nipping out the soft growing-points.

The cow crops the pasture evenly and systematically, leaving behind it a sward which the sheep can clean to the bone. In relation to its size the cow's food capacity is only slightly greater than the sheep's but, having the economic advantages of a larger unit, it can afford to accommodate coarser fodder. However, the sweep of its tongue must embrace 1½cwt (76·2kg) of grass each day, and the cow has neither time nor patience for anything that frustrates the steady rhythm of its grazing, be it short herbage, prickly weeds or wood shrubs. When the invading seeds get past the seedling stage the main defence against reversion is the goat, assisted in recent centuries by man.

The goat faces its task with a hero's equipment. It has the toughest mouth of all the ruminants and can consume with profit and pleasure such well-protected vegetable treasures as the bramble, briar, thistle and nettle. In proportion to its size the goat can eat more fodder each day than either the sheep or cow. Because it is a browsing animal, rather than a grazing one, and has a great ability to select the most digestible and nourishing portions of the plants it finds before it, coupled with a willingness to eat many plant types avoided by sheep and cows, its food intake is digested in the rumen more rapidly than that of its grass-eating companions; consequently it can manage a greater intake, and can provide itself with a better supply of nutrients, in certain environments, than sheep or cattle. This will be discussed more fully in Chapter 6, where it will be seen that, in fact, the rumen of the browsing goat is narrower than that of the grazing sheep or cow, in relation to body size.

Even in a temperate climate such as that of France, and on cultivated pastures, the goat has been shown to use 15 per cent more varieties of available pasture plants than either sheep or cattle. Vegetation which is exposed to great heat, drought or frost must protect itself from freezing or evaporation with a tough fibrous skin or texture. If sheep or cattle are to graze such pasture at all, they will benefit from the presence of goats to control its more fibrous elements. When conditions become extreme, the goat is left in sole possession.

The wild ancestors of the domesticated goat are *Capra aegagrus* of Persia and Asia Minor, *Capra falconeri* of the Himalayas, and *Capra prisca* of the Mediterranean basin. The common goat of most of Europe and Asia is derived from *Capra aegagrus*; the Kashmir and Cheghu goats from *Capra falconeri*; the Angora goat from a cross *aegagrus* and *falconeri*. The remains of the domesticated offspring of *Capra aegagrus* have been found in Early Stone Age deposits in Switzerland, and it is to be presumed that this is the only type to have reached northern Europe until modern times. But investigations in Egypt by Professor de Pia have shown a prehistoric dwarf goat replaced by domesticated derivations of the twisted-horned *Capra prisca* before the advent of modern derivatives of *Capra aegagrus*. It is probable that *Capra prisca* provides some of the ancestry of most modern goat stocks in the Mediterranean basin, and that this accounts for some of their distinctive characteristics.

Apart from fleece production, the goat is valued mainly for its meat and milk. As a milk producer the goat is inevitably more efficient where the available fodder is of such low quality that a cow can barely live. On the desert fringes of the Middle East the cow doesn't get a look in; milk supply is in the care of the Mamber goat and its relatives. The Mamber

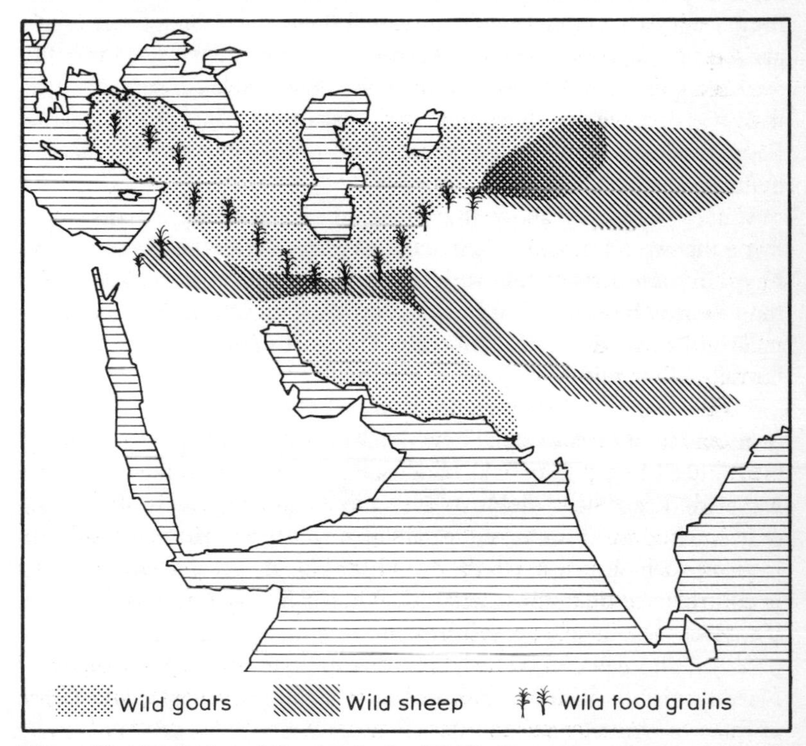

Fig. 1 The beginning of farming in the Middle East *c.* 10,000 BC.

is a large goat, weighing up to 120lb (54·4kg) with long black hair which is used mixed with wool for carpet making; she yields up to 5 pints (2·8 litres) a day when well fed.

The cow dairy business does not start until the quality of fodder is such that the cow can give 2 gallons (9·1 litres) of milk a day at the peak – that is, feed her calf and provide a gallon (4·5 litres) a day for sale. On a lower level of feeding than this, a goat can feed a kid and provide about 3 pints (1·7 litres) of milk at the peak to her owner (i.e. where the forage provides an energy level of about 11MJ ME/kg of dry matter). The advantage of the goat is further extended by her fecundity: two kids are normal in temperate climates; triplets and quads are common with well-fed goats in warm climates; one or more of these provides a valuable carcass at a fortnight old. But it is significant that in the countries where they are most extensively bred, the average yield of goats is just at this marginal line of about 3 pints at the peak, while suckling kids under control. Such is the figure given for the two main breeds of Indian goats, the Jumna Pari and the Beetal, though individuals of both breeds have proved capable of yields of 6 (3·3 litres) and 7 pints (3·9 litres) a day and over 100 gallons (454 litres) a year. Where lower yields are prevalent, as with the dappled Bar Bari goat of Delhi and the little black goat of Bengal and Assam, the breeds are dual-purpose, with a smaller, meatier body and higher fecundity.

Where yields rise above the 3-pint level, indicating a standard of living that would support a dairy cow, as they do with the Zariby goat of Egypt in some districts and with Granada and Murcian goats of Spain, the cow may be kept out of business by the difficulty in distributing its milk on bad roads in a hot climate. The goat delivers on the hoof, in household quantities; the cow is essentially a wholesaler.

The goats of Malta are to a great extent protected from the competition of dairy cattle. Their yields average 3 pints; but the 3 pints is produced by a 70lb (31·7kg) body, and the occasional full-sized goat has proved capable of yields of over 300 gallons (1,364 litres) a year.

It is when we come to the mountains where the pasture would do justice to a pedigree Ayrshire, but is at an angle at which no cow can graze, that we find substantial numbers of high-yielding milk goats. The Swiss Saanens, Toggenburgs and Chamoisées, also their counterparts on the eastern side of the Alps and the 'Telemark' goats of Norway, are all capable of yields of up to a gallon a day with a lactation of 8 to 10 months.

From these mountain breeds derive the substantial number of

scattered herds of high-yielding goats which are found about the industrial cities of northern Europe and in isolated rural communities in fertile agricultural districts. These herds owe their existence to the fact that, given the inherited capacity of milk production, the goat is a slightly more efficient converter of pasture into milk than the cow.

Preposterous at first sight, this claim has a clear enough theoretical basis, and has been exactly demonstrated in practice by the experimental work sponsored by the Department of Agriculture of the USA.

From the point of view of comparative costing, the maintenance ration of the cow or the goat is the amount of food each requires to keep her going, while converting a given quantity of digestible nutrients into milk. Each 100lb (45·4kg) of goat requires 1½ times as much maintenance ration per day as each 100lb of cow. But each 100lb of goat eats twice, or three times, as much digestible nutrients a day as each 100lb of cow. Consequently each 100lb of goat has more digestible nutrients available for conversion into milk each day than each 100lb of cow. For every 10-stone (63·5kg) bag of dairy cake fed to it, the average goat produces 2 gallons (9·1 litres) more milk than the average cow.

But the goats' milk cannot easily be produced in commercial quantities to compete in cost with that produced by cows on good land. A well-managed herd of goats will produce a yield of 200 gallons (909 litres) per head; management and cropping for thirty of them is as laborious as the work of a dairy farm with twenty cows, yielding 600 gallons (2,727 litres) a head. So the labour cost per gallon of goats' milk is roughly twice the labour cost per gallon of cows' milk. If the goats are sustained largely on natural waste vegetation, instead of crops, labour costs fall, feeding costs fall, and competition is possible. The man who doesn't charge himself anything for his own labour can usually produce goats' milk considerably more cheaply than anyone can sell or produce cows' milk. Moreover, while few Durham miners could buy or graze a cow, in the days when they were grossly underpaid, many of them were keen goatkeepers. The same principle holds good the world over; goats exist to cut the cost of living for many millions of hard-pressed families in cow-dairying country.

In France, in Norway and in the UK, special value is attached to goats'-milk cheese and in the United States as in this country, a specialized market for goats' milk has been developed based upon its higher digestibility and its value to sufferers from allergy to cows' milk.

Accurate figures for the world goat population do not exist but those produced by the Food and Agriculture Organisation (FAO) for the world and by Eurostat for Europe seem to be the best available. Although fluctuations occur in individual countries, the overall picture is one of increase. Between 1979 and 1991, world goat numbers rose from 460 million to 530 million, an increase of 15 per cent, greater than the rise in cows and sheep. Africa and Asia have 89 per cent of this total. Developing countries keep goats mainly for meat, the developed countries mainly for milk production, but there is a world trend to produce more goats' milk and less goat-meat. Greece has by far the most goats of all European countries, at over 5 million.

Apart from such dramatic movements in goat population, there is a steady response to world-wide and local fluctuations in the relative value of labour and food. Improved agricultural methods lower the price of food, and by increasing the value of land increase the labour involved in controlling goats; at the same time higher levels of land fertility can support the rivals of the goat in commercial dairying. While the goat is common in the Early Stone Age deposits of the Swiss lake settlements, it becomes rarer in the later ones; throughout history large-scale goatkeeping retreats before the advance of agriculture. On the other hand, scarcity of food forces interior land into production and demands the utmost use of available fodder; so we find a sharp rise in the goats' representation among farm stock in all countries during war and economic difficulty.

Broadly speaking, therefore, the goat's place in the world dairy business is primarily on land that is too poor, too hot or too steep to support dairy cattle; where such areas are extensive, a daily peak yield of 3 to 4 pints (1·7 to 2·3 litres) per goat provides a basis for commercial production, and in most cases is the maximum that the pasture will sustain. On better pastures, goats of good milking strain are capable of converting fodder to milk quite as efficiently as the best strains of dairy cattle; but as the labour cost per gallon of goats' milk is twice that for cows', this is a solid barrier to commercial goats'-milk production, which can be overcome only when goats' milk commands a higher price than cows' on the strength of its special merits or because of overall milk shortage, or where otherwise valueless and uncultivated land supplies the goats' feeding. Where labour is costless, as it often is when the goat is used for domestic supply, goats' milk is far cheaper than cows'.

In considering the goat as a meat producer, we must differentiate between kid-meat and goat-meat. Kid-meat is a by-product of the dairy goat, which is probably appreciated and utilized in every country in which goats are kept including Great Britain. Intrinsically it is in every way as valuable as veal, and is rather more versatile in the hands of a good chef. Goat-meat, when reasonably well produced, is in no way inferior to the general run of old ewe mutton, which was parsimoniously inflicted on the British dinner plate for fifteen years of war and 'austerity': but it cannot rise to the heights of prime beef or lamb and when it is produced in quantity it is produced for its own sake.

As beef cattle can rear a calf and acquire some degree of fatness on pastures too hot or poor to sustain dairy cattle, it follows that the breeding of goats for meat production is less extensive than the range of the dairy goat. In many areas where goats are the source of milk supply, beef and mutton dominate the menu. There remain regions of desert, dense jungle and high mountain pasturage where goats are both meat and milk to the inhabitants. Such is the case in Arabia, Syria, Iraq, in parts of Equatorial Africa and in the Himalayas. There are wider and more fertile regions where the value of cattle as draught beasts is such that it denies their use for meat. Throughout most of India the cow is sacred. Religion has endorsed the need to safeguard the producer of agricultural power units from the hunger of a chronically famine-stricken population. Goat takes the place of beef on the Indian menu; it is a happy arrangement: the firm, rather dry flesh of the goat takes well to currying, and is none the worse for its dryness in a land rich in relatively cheap vegetable oils. The wool of the sheep constitutes its main value in hot countries where its meat never reaches perfection; under tropical and sub-tropical conditions the goat is more palatable as well as more prolific.

In the cooler, developed countries, the fact that the goat carries its fat deposits within the body cavity rather than beneath the skin and among the fibres of the meat itself, as in sheep, has given the leanness of kid and goat meat a big boost in today's 'healthy eating' campaign. Those farming fleece-bearing breeds of goat, Cashmere, Cashgora and Angora, need to sell meat animals even more than dairy-goat farmers; this has led to the launch of the trade mark 'Capra' for Angora goat-meat of a certain quality produced in the UK. In Scotland, a marketing co-operative for goat-meat, usually known as chevon, is doing much to promote and market this fine product; it is the fruit of the labours of the Scottish Cashmere Producers' Association. Rearing costs, however, make it necessary to obtain 'gourmet' prices, akin to those for venison. As travel makes the world daily into an ever smaller place and mixes populations, the demand for kid and goat-meat in cooler countries grows, though it is unlikely ever to exceed a minority demand in temperate areas.

The skin, wool and hair of the goat strengthen its economic basis in many areas throughout the world. Skins are exported from India and Pakistan and African countries, notably Morocco, Somalia, Uganda and Nigeria. World skin production is around 345,000 tonnes. The most valuable are those strong but lightweight enough to produce glacé kid leather, and there are research programmes aimed at breeding goats with suitable quality skins for this product. Coarser skins are needed for strong footwear. Angora skins dressed with the hair on are used for rugs. The finest leathers, suedes, etc., are used for clothing and gloves.

The fine undercoat of the Cashmere (Kashmir) goat, which is 'farmed' by nomadic tribesmen above the 15,000-foot contour of the eastern Himalayas, has a superlative quality. The annual cashmere crop is 8–16oz (225–450g) per goat and is combed out with scrupulous care, the process taking eight to ten days. The majority of the world's production of cashmere is processed into the most luxurious, expensive cloth and knitwear in Scotland. The Cheghu goat of similar type of habitat in the western Himalayas is also used as a pack beast. In both cases the long outer coat is shorn and used for making tent-cloth, and mixed with other fibres for carpet-making.

Throughout the Middle East, Eastern Europe and India the hair of long-haired goats is used for weaving coarse cloths, making rugs and carpets and for packing mattresses. But the hair of the Angora goat, which originates from Turkey, is of a special quality, being a fine, lustrous silky fleece with an 8in (20cm) staple, which covers the animal to below its hocks, and weighs rather more than an average Blackface sheep fleece. This material, the mohair of commerce, has properties that combine high lustre, good receptivity to dyes and great durability. Plush and braid are less fashionable than they used to be and nylon has similar qualities, but mohair still meets a ready demand, providing the basis for furnishing velvets, for many of the better-quality fur fabrics, and for high fashion garments for both men and women, besides making a big contribution to the carpet and rug industry. Mohair is the *raison d'être* of several million goats in South Africa, in the dry states of North America and in Turkey. The Angora goat thrives best in a fairly warm climate with a rainfall of under 20in (51cm) per annum. The Australasian countries and most recently European member states, including the UK, are working to establish mohair, cashmere and cashgora industries, the anticipated climatic problems not having materialized.

From time to time, changes in economic conditions or agricultural techniques deprive the goat of its usefulness in an area where it has long played an important role. Swiftly and without fuss or publicity, the goat flocks dwindle and disappear. This has happened in Switzerland, where Alpine goat dairying can neither offer its labour force a wage, or way of life, to compete with the attractions of industrial employment, nor economize in labour by mechanizing its processes. So the goat flocks of the seven Swiss breeds are dwindling away at an accelerating rate. Actually, this reduction has now stopped, and a stabilization can be expected. Two reasons explain this: first, there is a good demand and a high price is given for goat cheese and kid-meat; second, the goat is recognized as having an important role in the care of the landscape in the Alpine region.

On the other hand, when we encounter an organized propaganda campaign against goats, prominent officials demanding the extermination of goats, and laws against goatkeeping, we can be sure that goats are indeed an economic proposition for their owners in the country concerned; and that their owners lack political power. In the countries of the East Mediterranean and Middle East such a situation was common in the day of colonial rule, and persists today wherever government is in the hands of aliens or more prosperous classes. Throughout this area at the crossroads of the continents the land has been subjected to man's ill-treatment more intensively and persistently than anywhere else on earth. For thousands of years every patch of watered soil has been cropped and cropped again, without any manurial treatment, until finally abandoned, exhausted, unseeded and naked to the elements. Starting from the lowland plains, man has continued the process up the mountain side; but on the slope the naked soil of abandoned fields is quickly washed away in erosive torrents that gouge great scars in the plains below. The primitive techniques of shifting cultivation still persist. A 1961 Lebanon survey showed half an acre (0·2ha) of abandoned farmland for every acre in cultivation; in Greece, Turkey or Syria much the same was true; 1¼ acres (0·5ha) of felled forest for each acre (0·4ha) under timber, said the Lebanon survey. Where have all the cedars gone? Down the long road of international commerce that has raped this land without ever fertilizing it.

The pastures that once fed a mixed herd of cows, sheep and goats, were taken for crops, cropped to exhaustion and abandoned naked to the harsh mercy of the climate. The wretched cloak of scrub, spattered about the eroded land today, is all the fodder for the grazing herds. Only goats can scrape a living from it.

In Greece, however, where the goat was said to have been responsible for deforestation in some parts, notably Crete, the goat population fell drastically after the Second World War until, by the early 1970s, it was felt that the numbers were below what was desirable. Accordingly, the Greek government sponsored a programme of expansion which, with the removal of the ceiling price on meat – including that of goats – in the middle of the decade, helped to build up numbers again.

It is nonsense to suppose that trees are the only or the best counter to erosion. On Mediterranean hills the combination of hot sun, low rainfall and thin soil is unfavourable to tree growth; a turf of deep-rooting grasses fortified by a scattering of drought-resistant evergreen shrubs is the best protection against erosion that soil and climate permit, over wide areas, and is quite probably more effective than forest cover, if only because it regenerates so quickly after being burnt. As a British experiment in Tanganyika proved, in grazing containing bush and scrub the goat frees more grass than it eats. Grass on stock pastures carries the larvae of internal parasitic worms. Worm larvae are liable to desiccation, so they keep mainly to the layer of dense vegetation close to the ground; the higher branches of bushes and shrubs are relatively free from larvae; in self-defence, the goat prefers them. Some years ago, in Venezuela, a cattle-ranching lobby prevailed on a cattle-ranching president to decree the extermination of goats over a vast area. The law still stands, but the goat is now back to its former strength and the government is engaged on a goat improvement scheme. In the goat's absence, the bush invaded the pasture, grass was smothered, cattle stock deteriorated and ranchers' profits fell.

Let us grant that goats wreak havoc in a young plantation and prevent natural regeneration in a felled forest; men seeking firewood and raw materials for Mediterranean do-it-yourself furniture have the same effect. Neither does much damage to mature timber. If trees must be planted in a place, goats must be excluded, and unauthorized men as well. But in the present context of human starvation and chronic malnutrition, prevalent throughout the East Mediterranean, there is a lot more need for goats than for trees. Goats' milk, goats' cheese and goat-meat are the main source of protein for the underfed and protein-starved majority in these countries. International aid, central planning, technical advice and political speeches are no substitute for protein in breaking the prevalent lethargy that balks development there. Though goat owners are almost all poor men, goats are, on the average, 50 to 75 per cent more profitable than cows for meat and milk production under East Mediterranean conditions. Sheep are no substitute; their milk is rich in fat, but vegetable oil is cheap; in milk-protein production they cannot compete with goats; mutton is a luxury. The attempt to replace goats by sheep, cows or trees in this area is merely a rich man's racket allied to bureaucratic laziness. It is easier for the bureaucrat to blame the goat than to pin the responsibility on the real culprit, to ensure that cropped land is properly manured, and sown out to grass, instead of being abandoned to weeds when its cropping capacity is exhausted.

For many years to come the goat must retain its pre-eminence in this part of the world, and in others with a comparable climate. But the present goat population is excessive. By improving the quality of the stock, fewer goats could make better use of the available grazing. A considerable percentage of the flocks at present are surplus males or aged females, kept not for the productivity, which is nil, but to provide social security insurance against the risks of disease, famine, and marriage in the family. To convert this surplus into dependable currency and sausage would be a kindness to man, land and goat. Better organization of the marketing and distribution of goat products would do away with the unproductive section of the flocks, and improve the health of a protein-starved human community.

Once goats are recognized and treated as playing a vital part in the national development plan, it is possible and necessary to tackle the other objection to their existence in this part of the world – their liability to carry Malta fever and TB. In fact goats, cows and pigs are all liable to infection with their own variety of brucella bacteria; all of them can pass the infection on to man. Man becomes partly immunized to the forms of brucellosis to which he is regularly exposed; but a world of travellers needs protection. Brucellosis can be eradicated from the Mediterranean goat as it is now being eradicated from the British cow. In the meantime, pasteurization of the milk of untested animals is a sure safeguard. The European Community has a legally enforceable scheme in operation to eradicate *Brucella melitensis* infection from sheep and goats.

The goat's place in world agriculture is primarily in the 'developing' countries, and its future depends on the direction that development takes. This refutation of the standard slanders against the goat is based on the reports and policy discussions of the headquarters of the Food and Agriculture Organization of the United Nations, the main channel of aid and advice to developing countries. For the first time for many centuries the goat has won some friends in high places. Who befriends the poor, befriends the goat. As humanity grows more humane, the goat grows in stature.

CHAPTER TWO

The Goat in Great Britain

Until the late eighteenth century, goats were a normal source of milk supply for cottagers throughout the country, and featured on every common in England. In the moorland and mountainous districts of the North of England, Scotland and Wales goats occupied a key position in the rural community.

On the hills sheep were kept mainly for their wool, much of which went to clothe their owners; cattle provided the cash income when sold for slaughter in the autumn, and perhaps a summer milk surplus for conversion into butter and cheese; goats bore the main brunt of domestic meat and milk supply. In these districts today you can hardly travel more than a few hundred yards without passing a spot whose local place name testifies to its former popularity with goats: on a modern Ordnance Survey map of the Highlands of Scotland you will find many place names with the 'gour' and 'gobhar' theme (Ardgour, Arinagour). The herds of goats you see here today are more likely to be for a combined purpose of cashmere production and pasture improvement, for goats and sheep have been found to make excellent complementary grazers, providing the stocking rate is right; the sheep eat the grass and clover, the goats consume the reeds and heather, allowing the grass to flourish.

On the thriving goat population of old there descended the two-edged sword of industrial revolution and improved agriculture. Improved agriculture brought in its wake the enclosure movement; the extensive common grazings, whereon the peasantry of England's cattle, ponies, sheep and goats overcrowded inferior grazings, were put under the plough and crops. This immense development of Britain's food-producing capacity took at times the dramatic form of high-powered robbery by the local landlord, accompanied by local disturbances and migration from the land. But in the main it was a long, peaceful process, whereby the peasant exchanged a penurious independence for a wage through which he shared in the increased productivity of his lost commons. In either case it involved a large drop in the goat population. Where the goat survived in the neighbourhood of enclosed land, it was a

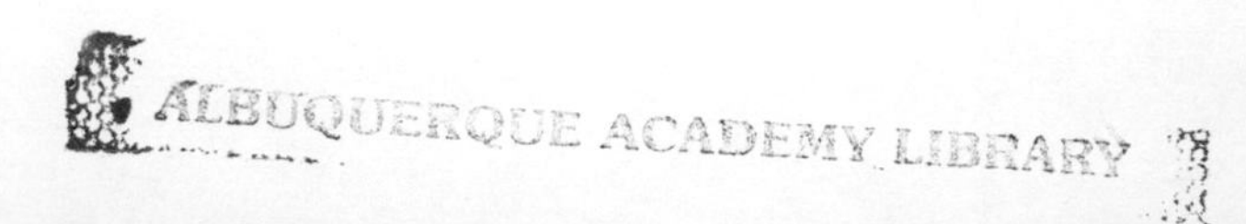

nuisance to the farmer of the fields it raided; and a nuisance to its owner, who was obliged to control it by constant herding or tethering. As the value of agricultural labour increased with the fertility of the fields, goatkeeping became more expensive; the practice of tethering inevitably increased the goat's liability to infestation with internal parasites, and reduced its productivity. We may be sure that then, as now, the most enthusiastic goatkeepers were the most awkward individualists in the area, and their goats the *casus belli* with their farmer neighbours. Where the enclosure movement took on its more dramatic forms, the goats were the spearhead of the underground resistance movement of the cottagers.

The face of agricultural advance was turned against the goat; and the goat chivvied the heels of advancing agriculture with its infuriating impudence and eternal disdain. To the undying credit of its nuisance value, an atavistic hatred of goat still lingers in the mind of England's farmers. Hated by the masters of the land, it declined steadily in the affections of the land's servants to the point of near-extinction.

Few tears need be spent on the departure of the goat from the common grazings of English lowlands. For the most part it was replaced by sheep and cattle that could produce milk and meat and wool more economically, and by crops that greatly increased England's stocking capacity for man and beast. The change was the first necessary step towards industrialization. Certainly neglect of the goat was carried too far; there was scarcely any development of breed type or methods of management for one hundred years of changing agriculture; much land that goats could use was left useless, and many an isolated cottager, whom goats' milk could have sustained, went milkless. But as the prime role of British agriculture became the feeding of the urban populations, the goat lost its place on England's better farming land.

The decline of goatkeeping in the mountainous and moorland areas was something of a national disaster. For the widely distributed inhabitants of these regions, the goat remains the most economical form of milk supply; the better use it makes of coarse fodder and its longer lactation have never failed to give it an advantage over contemporary cattle stocks. The reasons that the goats left the hills are therefore of more interest. They are perhaps best seen by a glance at the goats of the Scottish Highlands, where conditions are extreme and the extinction of the goat most dramatic.

Towards the end of the eighteenth century the Scottish Highlands supported approximately ten times the population that they do today;

the basic mode of living was a very low standard of subsistence agriculture, with crops of primitive bearded oats and 'bere' barley and a stock of cattle, ponies, goats and a few sheep. Cattle were bred primarily for beef and provided the only substantial cash crop and export of the area. The sheep were the old tan-faced breed, of little value as mutton makers, but producing wool of the modern Shetland type for domestic consumption. The cattle provided a little surplus milk during the summer, and barren cows and bullocks were also occasionally bled at this season. Otherwise domestic needs of meat and milk were met by the goats. From May to September the whole stock was driven away from the arable land about the croft houses, and herded from summer sheilings on the high hills by the young folk of the community. The cheese and butter made at the sheilings was the mainstay of the people's winter rations.

Improved agricultural methods began to filter into the Highland area from about 1750, and their most successful application took the form of breeding the improved Border sheep on the Highland hills. Looked at from a national point of view, this enterprise provided a meat surplus for export to the south about four times as great as that available from an area worked by the traditional methods. Looked at from the landlord's point of view, the Border shepherds offered rents five to ten times as great as those they could hope to extract from their clansmen. From the point of view of the mass of the Highland population, the new methods meant that a shepherd and his dog and six hundred sheep could profitably occupy an area that previously supported a crofting township of about a hundred souls – in brief, it meant mass unemployment.

In England the population was rapidly multiplying in the industrial cities; at the beginning of the nineteenth century new industries and the Napoleonic wars were draining the countryside of men and forcing up the need and demand for meat and wool. The woollen industry, a principal contributor to national prosperity, was cut off from its Spanish raw-wool market and faced with disaster. Many Highland landlords shelved their sentiments and their clansmen together and accepted the sheepmen's rents and a pat on the back for patriotism. The Highlanders were cleared from the more fertile glens and hill grazings, cattle, ponies, goats and all. Some were moved by threats, some were moved by force, and some were left to face the cold and unrelenting wind of economics.

Many of the evicted crofters were re-established on poor ground on the sea-shore, bereft of extensive hill grazings. Crops and the sea gave them a living; tiny patches fertilized with lime-rich shell sand and seaweed and cultivated by hand with immense labour stood between these families and starvation. This was no place for a goat, and any goats retained on holdings of this kind would have to be tethered to preserve the crops. As we shall see in a later chapter, tethering undermines the goat's natural ability to avoid parasitic worm infestation, and almost invariably results in reduced productivity. Cows, more easily warded off crops and producing more milk per man-hour of herding under these circumstances, ousted the goat from favour.

During the more violent clearances, the greater part of the goat population must have been left to run wild on the hills. Domesticated habits always sit lightly on the goat, even the most inbred and highly selected of modern dairy strains being perfectly willing to turn feral* if given a suitable environment and opportunity. There must originally have been a great stock of 'wild' goats of this kind on the newly acquired sheep ranges. But goats commence their breeding season when the hours of daylight decrease and the hours of darkness increase at a certain critical speed. This critical rate of change is, of course, reached earlier in the year the nearer you go to the land of the Midnight Sun. Over the Highlands of Scotland the breeding season starts in July and August, and the kids are born in January and February to the feral goats – and only the very toughest and luckiest can survive the inhospitable welcome the Highlands offer at that season. Moreover, the goat does not like the wet, and in a climate such as that of the Scottish Highlands the feral goat population is very much limited to the number of dry beds available on a wet night. Feral goat communities are still widely distributed in the Highlands of Scotland, but the numerical strength of each is fairly small and static.

Over wide areas of the Highlands and islands the crofters were not ordered or forced to make way for the sheepmen. Whether the slow attrition of poverty, hunger and discomfort was a kindlier fate than that suffered by the victims of the great clearances is doubtful; the result was very similar. Looking down on the ruins of abandoned croft houses standing among the head-high brackens that cover the old arable land,

*The term 'feral' is used here and in subsequent chapters to emphasize the fact that 'wild' goats in Britain are not 'wild' in the sense that red deer are 'wild' – they are, as the botanist would say, 'escapes'. The distinction is of some importance, for the survival of 'wild' goats is not altogether dependent on their adaptation to their environment, but partly on a continuous supply of 'escapes'. Consequently the existence of these 'wild' goats is not a proof that goats can thrive under wholly natural conditions in this country.

Plate 1 Galloway Wild Goat with male kid.

it is hard to tell today whether the desert was created by sudden violence or perennial poverty. The uncleared crofts collapsed through the interaction of two forces. The attraction of better-paid jobs in the south and overseas drew away the young and enterprising; the call grew clearer and more insistent as communications improved and the tales of successful emigrants multiplied. Those who remained behind were attracted by the relatively easy money that could be made, temporarily, by adopting the sheepmen's methods and marketing mutton and wool for a living. Subsistence agriculture is impervious to the ebb and flow of market prices; a pound of steak is a good meal whatever the price of store cattle. But once the peasant has gambled his fate on filling his pocket before his belly, he is at the mercy of the market. The market kills and blesses by turns, but cannot bless those it has once killed.

The goats on the uncleared crofts were directly affected by the call of emigration which drew away the human population and, in a different way, by the advent of sheep breeding and the change-over from subsistence agriculture to a crop-marketing economy. The tendency of the domesticated goat to turn feral, which is examined in detail in a later chapter, evinces itself at two seasons – at the onset of the breeding season in August–September; and at kidding time, if there is adequate natural feeding available. Kidding time with domesticated goats in the Highlands is best arranged to take place in March and April. Thus the tendency to tern feral, which requires careful and constant herding to prevent, occurs at seed time and harvest when there is a minimum of labour available to prevent it. The fall in human population released the goats to freedom and its perils.

The advent of sheep breeding in itself should scarcely have affected the goat population, for goats are good friends to the sheep and save many from death in bramble thickets and from falls from the cliffs. Yet there is a fable still prevalent among the sheep fraternity that goats on a sheep grazing will interbreed with the sheep and ruin the stock. The ram will serve the she-goat and the he-goat will serve the ewe; but only if frustrated in search of their own kind; very occasionally, as a rare phenomenon, the service bears fruit; there is a possibility that occasionally a ewe so served may go barren after an early abortion. On this gossamer of fact, which might support the delicate interest of a biologist, shepherds have hung an axe of execution. The goats on the Isle of Ulva (off the coast of Argyllshire) were slaughtered under this pretext some years ago. My own landlord deprecated my goatkeeping activities for the same reason. In the days when science and shepherds were strangers, many of the Highland goats went innocent to the slaughter.

The goat is by its nature the symbol and mascot of subsistence agriculture. It is first and foremost a household provider, and in this role its useful characteristics find their fullest expression. With the decline of subsistence agriculture the demand for goats declined. Under the pressure of a crop-marketing economy, the Highland crofters turned to cattle with some dairy qualities; with these they could go through the motions of eating their cake and having it. The calves were starved of milk to feed their owners, who then cashed in on their stunted maturity. So the desperate hunt for cash deteriorated the cattle stocks, diminished the net income from the area and hastened the process of devastation.

When first the sheep invaded the Highland hills they throve on fresh pastures, almost free of internal parasites, and wintered on abandoned croft land still fertile from centuries of human labour. They sold to a roaring trade in a hungry market. The honeymoon was brief. By the later half of the nineteenth century, sheep farming paid so poorly that the Highland landlords were clearing off the sheep to give unfettered scope to the stalking tenant and his quarry. If sheep were a nuisance to the stalker, goats, the most keen-sighted of all our quadrupeds, were worse, their snorting whistle of alarm at the sight of a stranger carries farther than the sheep's. Feral goats still suffer from the stalker's rifle for this sin; domestic goats are still banned on some stalking estates. In the heyday of stalking, the cold war of today was too hot for a lot of goats. Here again the goats were preyed on by a fable: many years ago, on the shores of Loch Morar, a billy goat was run with the crofter's cows; this one had not the popular excuse of preventing contagious abortion; he was there to frighten away the red deer hinds. The idea is plausible: the stink of billy within a mile upwind would mask from the hind the more delicate aroma of an approaching stalker and undermine her sense of security; so she would move out of range – and a rank billy has some range. If the fable were popular, it must have killed some goats. I call it a fable on the authority of Dr Fraser Darling's observations of the deer and feral goats of An Teallach (*A Herd of Red Deer*, F. Fraser Darling, Oxford University Press). It seems the hind's sense of smell is too selective, and its nature too inquisitive, for it to be so deterred.

With the exception of the influence of deer stalking, this tale of the decline of goatkeeping in the Highlands of Scotland covers the causes of its more gradual and piecemeal disappearance from the mountain districts of the North of England and of Wales. The change from subsistence agriculture to a cash economy; depopulation of hill districts; the tendency of the goat to turn feral if allowed free range without adequate control; and the goat's sensitivity to infestation with internal parasites when its grazing is restricted – these are the factors which have driven the goats from the hills of Britain. It is to be noted that while this exile from the hills is due in part to basic and irreversible changes in human society, it is due in part also to problems of management and husbandry which are by no means insoluble.

So far we have dealt with the goat population of Britain of one to two hundred years ago. The absence of accurate statistics of goat distribution at this period will not surprise the reader, who may be content to accept the evidence offered by local place names, songs and stories, and the statements of such authorities as Bewick, as to the approximate numerical strength of the goat. But now we have to admit that in the late

1970s, when the hens of Britain are numbered with annual accuracy, there is no authority who can state the number of domesticated goats in Britain to the nearest five thousand with any statistical justification. The Ministry of Agriculture and the Department of Agriculture for Scotland hold statistics based on agricultural returns; that is, they can state the number of goats for which rations were drawn during the period of feeding-stuffs rationing. This represents anything between 10 and 80 per cent of the actual goat population, according to district. The British Goat Society and the Department of Agriculture for Scotland can also offer figures of the numbers of goats in each district which were served by subsidized stud males. While these figures cannot form the basis of any accurate estimate of total population, they are useful in indicating the ebb and flow of interest in goat breeding from time to time and place to place from the 1930s onwards. For the rest, our information must come from show reports, memoirs and articles written by the older generation of goat enthusiasts.

From the end of the nineteenth century until 1940 the main goatkeeping districts were in the North of England – Yorkshire, Durham, Lancashire, Northumberland, Cumberland – and parts of Wales, forming a natural geographical and economic area featuring extensive moorland and poverty-stricken industrial populations. The goats distributed their services almost equally between rural cottagers and such ill-paid industrial workers as the miners and railwaymen were then.

In the rest of England goats were more sparsely distributed, with local strongholds in areas of unspecialized smallholdings, such as Essex, and frequently associated with industrial villages.

The goat of those days was a shaggy creature of nondescript colour, yielding up to 4 pints (2·3 litres) a day. There were two main types – the Old English, with horns sweeping up, back and outward in a smooth curve, rather short on the leg, with a long lactation, giving milk with a butter-fat content of 4 to 5 per cent. The Irish was more popular in the hilly districts; and the type survives still in the west of Ireland and in many feral herds in England, Scotland and Wales. In the Irish type the typical horns rise straight and parallel from the brow, turning outward and a little back at the top in billies, remaining straight, pointed and business-like in the females. Leggier, with a shorter lactation, lower butterfat percentage, and somewhat lower yield, the Irish goats were annually imported and distributed through the hill districts of Britain in nomadic droves from which the milkers were sold as they kidded. Until

Fig. 2 Feral goat heads. The old billy on the left is of the type commonly found on the small islands off the Scottish coast and in the Border hills; the type is probably native or Scandinavian in origin, but local traditions describe them as survivors of the Spanish Armada. The head on the right is that of a typical Irish billy such as the Irish goatherds used to import annually into the hill districts.

the First World War the Irish goatherd, shouting picturesque advertisement of his wares, squirting great jets of milk from his freshened nannies up the main street, was a regular harbinger of spring in the mountain villages.

Some of the feral herds along the west coast of Scotland are pure white. Local tradition attributes their origin to the ships of the defeated Spanish Armada which sought here a refuge they didn't receive. Certainly, most large galleons of Elizabethan days carried goats as a source of fresh milk and meat, but there is little Spanish about the type of these feral herds today. The Spanish dairy breeds are all coloured, and such of them as are horned carry the short twisted horn of *Capra prisca*. However, the Netherlands were part of the Spanish Empire in those days, and the white goat of Swiss Saanen type has a long history there. Perhaps these were the first Saanen importations. It is equally likely that they owe their origin to Scandinavian seafarers who frequented these shores, in whose homeland the white 'Telemark' goat has long been popular. The sea route to the Western Isles was assuredly more hospitable than the land route until the late eighteenth century. Many of these feral flocks exist on small islands. But goats are bad swimmers; the goats of Ulva were exterminated by being driven out on to a tidal reef when the tide was rising. The ability to swim a hundred yards would have saved them. So it is highly improbable that these island goats swam to shore from wrecked ships. The prevalence of the tradition that they did so suggests that their origin is wrapped in mystery and antiquity. If their existence were due to the obviously sensible practice of sending dry stock and males to uninhabited islands for the summer – to save herding them from the crops – then the mystery would not exist, and the colour of the goats would not be so prevalently white. It is tempting and not unreasonable to suggest that the Viking longboats, which carried cattle to Greenland in AD 1000 and pirated West Highland waters for centuries, may have carried the white goat of Norway to the Western Isles and their islet strongholds. It is altogether appropriate to believe that the Vikings sustained their heroics on a diet of goats' milk and kid. In any case, there can be little doubt that the native goat stock along the whole seaboard of Britain was liberally mixed with 'ship goats' from abroad.

The relatively poor performance of the native British goat stock was not, then, primarily due to inferior origins, but to lack of a coherent breeding policy among British goatkeepers. In most parts of Europe, native goat stocks have long been developed to a high level of excellence, especially where a measure of inbreeding and coherent breeding policy has been enforced by the isolation and essential discipline of tightly knit mountain communities, such as those of the Alps and Scandinavia. In Spain the guiding hand of the centuries-old Sheepmasters' Guild has aided the production of a number of distinctive, long-established and highly efficient breeds. There is evidence that exposure to low temperature is of assistance in developing the goat's capacity for food intake on which the efficiency and economy of goats'-milk production very largely depends. Lack of the stimulus has perhaps been a handicap to British and Irish breeders. Whatever the cause, the dawn of the twentieth century found Britain with a goat population not only small in numbers, but relatively low in quality.

The first steps to improve Britain's stock were taken by the founders of her Indian Empire. The ships carrying the cargoes of the East India Company bore homeward also executives of the Company who had become acquainted with the virtues of goats' milk and its giver during their sojourn in the East. For the sustenance of passengers *en route* it was customary for these ships to take with them a few goats from India, and to pick up replacement milkers at Suez and Malta. On arrival in England, these unfamiliar animals found their way to zoos or to the home of a retired Anglo-Indian, who had acquired a taste for goats' milk, often allied with an impaired digestion which would not function without it.

These new goatkeepers were men of enterprise and vision with that Victorian faith in human progress which invested hopefully and earnestly in all manner of speculations, animal, vegetable and mineral. Under the aegis of the retired Anglo-Indians and their friends, the goats of the Orient were crossed with each other and with the native stock. Swiss goats were added to the collection, and in 1879 the British Goat Society was founded. Even today, the management of the British Goat Society and of the leading pedigree herds remains substantially in the hands of the spiritual and real heirs of its founders. These were not men of the land: they might have had one foot in the paddock, but the other was assuredly in the city. Consequently the modern dairy goat has not been developed in the farmyard like the cow, sheep and pig, but in the paddock and the back garden where, by garden-scale methods, have been grown such of the foods of Britain's modern dairy goat as the corn merchant has not provided. The virtues of the 'improved' goat have not been selected for their value in the general context of British farming, but for their value in these highly specialized surroundings. Pedigree goats have no more been bred to take their place in the mainstream of British agriculture than have pedigree Siamese cats.

It is not intended to belittle the achievements of the pioneer and present breeders of the modern dairy goat, but to explain why their brilliant achievements have had so little commercial success or practical application. The magnitude of their achievement is hard to exaggerate. From a mongrel hotch-potch of goats with yields of 50–70 gallons (227–318 litres) a year they have built up seven distinct breeds with annual yields averaging 150 gallons (682 litres) under fair management and 200–400 gallons (909–1,818 litres) in the leading herds. They have, moreover, succeeded in tying conformation to production so closely that few adult winners in the exhibition classes are absent from the prize lists in the milking competitions. Lactation periods have been extended to cover two years, and productive life up to ten years. All this has been achieved in about a century with the help of but a few dozen imported goats.

Before these achievements, and the productive potential they represent, could (or can) be developed into any great asset to national food production, two steps were necessary: first, to impress the improved type on the native stock; second, to discover the conditions and methods whereby the potential could be realized on any considerable scale within the context of British agriculture. We may win our battles on the playing-fields of Eton, but we cannot feed our millions on the produce of pony paddocks.

The U-boat campaign of the First World War gave the new goat its first opportunity. But numbers were still too small, and type insufficiently well established, for it to make any sizeable contribution to the national larder. Goatkeeping in general received a fillip, and more might have been done to improve the type of the national goat stock had agriculture been subject to any but the crudest organization. In the upshot the new goat found an extended patronage which survived the peace in a few more paddocks and back gardens.

The agricultural slump that succeeded the First World War slowed the development of the new goat, but didn't stop it. The general economic depression which followed produced the first notable step forward: on the one hand goatkeeping increased in importance in industrial villages and on the outskirts of towns in the depressed areas, to provide occupation for the unemployed and cheaper food for the hungry; on the other, official recognition of the new goat's value took the form of the Stud Goat Scheme, subsidized by the Ministry of Agriculture and operated by the British Goat Society, whereby the services of improved males were made available to the goats of cottagers and smallholders at reduced fees. A parallel scheme operated in Scotland under the auspices of the Department of Agriculture. By the outbreak of the Second World War the ever-growing popularity of this scheme and the steady progress of the leading pedigree breeders had established a nucleus of about 5,000 pedigree goats, with a further 15,000 showing some degree of improvement.

With this nucleus, the impetus of the national need and the whole agricultural organization of the country firmly under government control, the scene was set for the new goat to take the stage.

Male goat licensing would have brought about quick improvement in type; but the Ministry of Agriculture was already embarrassed with manpower shortages. Knowledge of goats and knowledge of agriculture were so seldom to be found together in the one head, that the War Agricultural Executive Committees did nothing of consequence to utilize the goat's potentialities to meet the national need. The ever-increasing acreage of cut-over woodland, which goats might have converted into low-cost milk, remained as a verminous fountain of weed seed to poison surrounding fields. With little more than 30,000

Plate 2 R.46 Bitterne Tessa*. British Goat. Her remarkable fourth lactation was 12,393lb (5626·42kg) in 1,061 days, butter fat average 3·84 per cent. Photographed when running through for the second time. Twice winner of the Malpas Melba Trophy. Breeder, the late Miss Barnaby. Owner, Mrs J. Laing.

goats on the Ministry of Agriculture's records, the Ministry of Food reached the understandable conclusion that goat dairy produce was best treated as an insignificant anomaly. The British Goat Society, somewhat enfeebled by the absence of many of its most vigorous members on active service, fought a gallant uphill fight for the elementary privileges and food allocations that were automatically accorded to all productive livestock except goats. The goat's real potentialities for national service were never so much as pleaded.

The consequences for the long-term future of goatkeeping in Britain were, in some ways, catastrophic. Goatkeeping expanded, as it always must when dairy produce is in short supply. But it expanded in the artificial economy of the subsidy and price-control system, almost totally divorced from its true sphere of usefulness and economic justification. The national Goat Stock improved in type and productivity under conditions which alienated it still farther from British farming.

The value of the goat lies primarily in its ability to convert to milk herbage that no other animal can utilize. Its suitability as the household dairy supplier enables it to cut the cost of milk distribution in some areas, and to utilize wastes available in quantities too small to be otherwise useful. Under the subsidy system, cows' milk was made available to all at prices below the cost of production and distribution; the welfare foods service made it available free or at a nominal price to a considerable section of the population, including children. Unsubsidized goats' milk became overnight a luxury food.

Goatkeeping declined wherever its previous justification had been economy. It was almost completely ousted from the industrial villages, retreating before the joint forces of subsidized cows' milk and rising wages. War work for women and labour shortages helped to drive the goat from cottages and smallholdings within the range of a dairy van. In general there was a retreat from the hill districts and the less well-to-do counties of the North of England and Wales, from all the strongholds where the first principle of goat management was maximum production for minimum cost in feeding; war economy spelt a new role for the goat as the supplier to a legal black market in dairy produce: unsubsidized, goats' milk, cream and cheese were also free of price control. The focal centre of the goat population shifted south to the home counties and the heart of the national black market.

No stigma attaches to the operations of goatkeepers under these circumstances. They did that which government regulations permitted and obliged them to do. A legal black market in luxury products has a definite social and economic value under conditions of war and economic stringency. Goat dairy produce and the other items that resided under the national counter solaced the jaded worker, met the unscheduled need and provided a valuable cornerstone to national morale. Though its potentialities were not fully exploited, the goat served a useful wartime purpose; it was not Austerity Britain that suffered by the arrangement so much as the goat.

The cost of milk production under these circumstances was not a matter of prime importance: ice-cream manufacturers were willing to pay 10s (50p) per gallon (4·5 litres) for goats' milk; fresh goats' cream sold wholesale at 10s per pint (0·6 litres). Animal feeding stuffs, though often scarce, were price-controlled and heavily subsidized – and this was a subsidy from which goatkeepers did benefit. Labour was the essential of goats'-milk production in shortest supply and steepest in price. The obvious and most tempting policy for the goatkeeper to adopt was to secure the maximum production for labour expended, irrespective of costs in general and the cost of feeding in particular. In practice this led to the maximum possible use of concentrates to obtain the highest possible yields per head.

Between 1939 and 1949 animal feeding stuffs were rationed. For much of the period the goat's concentrate ration was limited to about 1lb (454g) per day for a milker giving over six pints (3·3 litres), ½lb (227g) for lesser producers and a very small allowance for kids, to encourage early weaning. Towards the end of official rationing, admittedly, it was easy to obtain extra coupons; throughout the duration of rationing the local farmer would let the goatkeeper have the odd bag of oats on the quiet; but lack of concentrates, and especially lack of protein concentrate in winter, was a real headache for the goatkeeper for most of that ten years. Pre-war goatkeepers complained bitterly that they were unable to keep their goats in the style to which they were accustomed; newcomers to goatkeeping accepted the low level of concentrate feeding more readily, but everyone attempted to compensate by feeding additional unrationed bulk food of various kinds.

During these years of austerity, there was a good demand for goats; breeding was far from selective; flock numbers doubled; and the average productivity of pedigree goats increased by 11 per cent, with a 6 per cent increase in butterfat. In the next five years, the supply of concentrates was unlimited; the demand for goats decreased; breeding became highly selective; and flock numbers were halved. Yet average yield fell by 14 per cent, with a 12 per cent fall in butterfat. Had the breeders of that period retained their austerity feeding methods and used the worst available males, they could not have achieved such a catastrophic destruction of breed quality in so short a time.

Imagine an exemplary disaster like this befalling any other section of British livestock: the breeders would be in dismay, the causes investigated, the methods reformed. No such disquiet disturbed the goat fraternity (sorority?): they trotted on down the Gadarene slope for another five or six years before a lonely voice was heard to remark that 'things ain't what they used to be'. In 1965, with yields now 20 per cent below the 1948/9 peak, and still falling, some misgiving became generally apparent.

The technical reasons for the collapse of productivity may be found in Chapter 6, but the reasons are not all technical. The fact is that from 1950 on, productivity had no great economic value for a majority of pedigree goat breeders. With cows' milk derationed and cows' cream legally on sale, goats'-milk sales ceased to be profitable; much goats' milk was diverted into the feeding bucket, for pigs, calves, chickens, pedigree cats and puppies, mink, silver fox – any consumer which provided some return without demanding an all-year-round supply, hygienic dairying or salesmanship. The profitability of such goatkeeping, if any, came to depend largely on sales of pedigree goats to other breeders, a good show record being worth more than a high yield.

The recent spectacular rise and fall in the popularity of goatkeeping in Britain is revealed by the figures in Table 13 (p. 156). There are several reasons for this. In some areas cows'-milk deliveries have now been discontinued, forcing those with facilities to consider keeping goats. Families disenchanted with city life settle in country cottages with anything from half an acre (0·2ha) to several acres. There is a general awareness of the higher level of nutrition and palatability derived from fresh, home-grown foods. Agricultural shows often include goat classes, in which beautifully turned-out animals with well-filled udders catch the public eye. So the household goat, with her modest housing requirements and lavish milk yield in relation to her small stature, and her ability to make use of weedy corners and all the extras from the vegetable garden, has proved herself invaluable to those attempting some measure of self-sufficiency.

In addition to milk and its derivatives, male and any surplus female kids, milk-fed when milk yields are at peak, kill out well at three-to-five months for the table or deep freeze. The NPK rating (nitrogen, phosphorus and potassium content) of goat dung is exceptionally high – another asset, ensuring, through the compost heap, fertility for successive crops. Local goat clubs are doing good work finding reliable stock and starting these new goatkeepers, many of whom have no previous experience of dairy animals, on the right lines.

The self-sufficiency boom was, however, not the only reason for the growth in the numbers of dairy goats. Cows'-milk quotas, overproduction of foods of traditional varieties, and the interest of widely travelled consumers in 'foreign' delicacies such as goat cheeses, forced

Plate 3 CH: Peggysmill Rachel Q*BR.CH. BT5774P British Toggenburg goat with consistently good butterfats. S. Peggysmill Bluster BT5584. D. Westcliffe Miranda BT5267P. Consistent winner in milking competitions. Classic wedge-shape conformation. Owner/breeder, Miss Rosemary Banks. (Photograph: BGS Year Book 1976)

farmers to look elsewhere to boost falling incomes. 'Diversification' became the 'in' word in farming circles. Among the guest-houses, golf courses, equestrian centres, school nature trails, changing demand was turning dairy farmers into goat farmers, who were 'adding value' to their milk by producing yoghurts and cheeses. The Goat Producers' Association (Great Britain) Ltd was formed in 1984 to cater for the needs of these new professionals, and made huge strides in ensuring that the public were presented with hygienic produce, properly and attractively packaged. Feed firms, milking machine manufacturers and all other makers of farming sundries joined the boom, and the Goat Veterinary Society, formed in 1979, went from strength to strength.

The Royal Agricultural Society of England displayed goat farming at the National Agricultural Centre in Warwickshire, and research projects concerned with goat farming appeared in the annual reports of research institutes, both north and south of the border. By this time the Ministry of Agriculture's ADAS staff included goat advisers. In the mid 1980s it seemed that goat dairy farming was here to stay, but there were still some hard lessons to be learned. A feasibility study of the UK dairy goat industry, commissioned by Food From Britain and published in 1990, pointed to the fragmented state of the industry and its lack of co-ordination.

Marketing co-operatives have been tried but have almost universally failed. It has been said, perhaps with truth, 'Co-operative, British farmers are not!' Many of the new hopefuls soon discovered that, without the Milk Marketing Board tanker to collect the milk each day, as with their cows' milk, the strain of being producer, processor and marketing man all rolled into one was just too much. Sadly, the Goat Producers' Association was wound up through lack of funds in 1990.

Nevertheless, those with the ability to survive the pressures and to do well with goats have a good future, for consumer demand is high, evidence for this being the shelf space given by supermarkets to goats' milk and cheeses. The British Goat Society attempted to fill the gap left by the demise of the Goat Producers' Association by appointing a Commercial Liaison Officer, an experienced goat farmer, willing and able to give sound advice. ADAS, the Scottish Agricultural Colleges and the Goat Advisory Bureau in Somerset are also there to help the aspiring.

The cause of the drop in dairy goat registrations, seen in the column for 1990 in Table 13, is by no means entirely due to the closing of professional goat dairies. The self-sufficiency ideal faded with the memory of a TV sitcom portraying it. Soaring house prices and mortgages, a bad recession and a case of listeriosis in a woman who had eaten goat cheese – hyped up beyond all reason by the media – are some of the reasons for the decline in domestic goatkeeping. Another factor, the effects of which may well be far-reaching, is the advent of legal control of the production of goats' milk for sale for human consumption. Milking in the goat house, cooling and cartoning the milk at the kitchen sink then selling it, for so long part of the goatkeeping scene, is being swept away by the hygienic brooms of national and European government (see Chapter 11).

'The Goat in Britain' no longer implies 'The Dairy Goat in Britain'. For the long-forgotten options of cashmere and mohair production have been spectacularly brought back with the importation and careful breeding-up of cashmere-bearing and Angora goats, and the development of a market for their fleeces (see Chapter 14). The Boer goat too, a South African breed developed to massive proportions to supply meat, has been imported into the UK (see Chapter 13).

A further revival from bygone days is the use of the goat as a harness animal. The public displays given by members of the Harness Goat Society, their goats and their vehicles, always draw and delight the crowds (see Chapter 18).

No survey of goats in Britain could be complete without the mention of three further breeds: the English, the Bagot and the Pygmy.

The English goat is an attempt to recall the hardy milk and meat producers of the cottagers of the last century, mentioned at the beginning of this chapter, before the influx of Swiss blood so drastically altered them. Typically greyish-brown with a black dorsal stripe and no tassels or Swiss markings, it is not colour but characteristics which count most. The English Goat Breeders' Association inspect all animals prior to registration.

The Bagot goat, so called because of its close association with the Bagot family at Blithfield Park in Staffordshire, has longish hair, coloured black on the head and forequarters and white on the rest of the body. It is non-productive, being kept out of interest as a rare breed; it is registered by the Rare Breed Survival Trust. Its origin is disputed: it bears a resemblance to the Schwarzhal goat of Switzerland and also to many of the feral goats of Wales.

The Pygmy is one of the most popular breeds in Britain. Based on various African pygmy breeds, the height of even the male must not exceed 56cm to the top of the withers. Kept as non-productive companion animals, these goats are registered by the Pygmy Goat Club.

CHAPTER THREE

The Prospects for Goat Farming

The previous chapter explained that it is now possible to farm goats for milk, mohair, cashmere, or meat. Of these meat is more often a by-product of dairy or fibre farming, and it must be said that a more rapid return will probably be obtained from milk production than from either type of fleece.

Dairy farming

Due both to the high quality of the products already on the market, and to the legislation controlling it, food production today cannot be regarded as a hobby. Unless you are 100 per cent committed to thinking of yourself as part of the food industry, with all that it entails, then forget the whole idea. Otherwise you will end up bankrupt, or in serious trouble with the law, or both! Much has been written, and all of it true, on the impossibility of farming the goats, producing the milk, processing it, marketing it, advertising it, delivering it – while in the spare time you do the books, all on your own. You must have help, and that help has got to be well paid.

As with any other job, it is not possible to do it really well unless it pleases, so it is best to try working with someone else's goats first, before getting your own herd; you may love them or hate them! The pages of this book will give guidance on how to look after the herd, but there is no substitute for practical experience.

Next, a market has to be established. This is difficult to do when you have nothing to sell yet, but not half so bad as having hungry goats, gallons of milk, and no market for it. Talking to cheese wholesalers and joining the Specialist Cheesemakers' Association may prove very helpful.

Financial capital is next on the list, and requirements will vary greatly, depending on whether buildings can be adapted, yards or fields are already fenced, there is already a bulk tank and maybe a pasteurizer, etc. Variable costs, for such supplies as feed, must be allowed for, and to milk 100 animals is about the least from which a living can be made. Most enterprises run at a loss for the first couple of years, as it takes time to establish a herd in which every goat gives at least 1,408 pints (800 litres) of milk a year; this is regarded as the lowest commercially viable yield. Chapter 8 gives a guide to deciding which breed to choose.

The new food laws cover temperature of storage and delivery, ripe soft cheeses being particularly targeted. Produce rooms and refrigerated delivery vans will eat into capital. Many people attempt only part of the total operation, either they sell milk for processing or process bought-in milk. While this does make life easier, it can be disastrous to be dependent on the success of someone else's business.

The feasibility study of the UK Dairy Goat Industry, mentioned earlier, produced some figures for costs and margins which are useful as a guide (see p. 40). It will be evident from them that high-yielding goats are necessary to a profitable enterprise. Making the milk into cheeses, yoghurts, ice-cream, etc., might increase profits. It must be understood that these figures apply to a farming situation. Where hay has to be bought in, for instance, forage costs could double those given. Also it is assumed that there is a market at the price quoted for all the milk produced.

Angora goat farming

Compared to dairy goats, mohair production requires a lower investment in money and time; the returns are also likely to be less, and to fluctuate with the demand, or lack of it, of the fashion trade.

A two-stage feasibility study, 'Prospects for UK Mohair Production' and 'Mohair Marketing in the UK', commissioned by the British Angora Goat Society and Food From Britain, was published in 1988. The conclusion reached was that Angora goat farming could be more profitable than sheep farming, with which it has certain similarities, but that this potential rested on a number of factors – mohair prices, the sale of goat-meat, the production of a sufficient amount of top quality fibre and the conviction of the mohair processors that Angora goat farming was being taken seriously in this country.

As with dairy-goats, the quality and productivity of the livestock is of key importance, and huge improvements have been achieved by breeders in the last few years. After a depressing period of low prices and lack of buyer interest in mohair sales in South Africa (the world

Fixed costs per goat in a herd of 100:

		£
Machinery	Maintenance and repairs	2
Depreciation	Herd	20
	Equipment	20
	Buildings	5
Insurance		4
		£51

Variable costs:

	£
Concentrates	60
Forage	35
Straw	10
Veterinary & Medicines	10
Herd replacer rearing costs	5
Chemicals	3
Heating/electricity	5
Transport	5
	£133

Financial return per milking goat in herds of different performance:

	Low	*Average*	*High*
Milk yield per year in pints (litres)	1,056 (600)	1,408 (800)	2,112 (1,200)
Milk price per pint (litre), pence	17 (30)	17 (30)	17·6 (31)
Milk value (£)	180	240	372
Progeny value (assuming 1·5 kids per dam, 15% for replacements)	£30	30	30
Culled goats value	£3	3	3
Less cost of replacements	£12	12	12
Total output (i.e. income)	£201	261	393
Variable costs (more for food in higher yielding herd)	£ 133	£ 134	£ 155
Gross margin	68	118	238
Gross margin/acre (ha) 3·2 (8) goats	220 (544)	382 (944)	770 (1904)
Fixed costs	51	51	51
Profit per goat	17	67	187

Figures prepared by Alan Mowlem, Goat Advisory Bureau, 1990

centre), the British Angora Goat Society's newsletter for summer 1992 was able to report some improvement. Prices (translated into sterling) ranged from a low of 82p/lb (£1.82/kg) for 'strong adult', the coarsest fibre, to a high of £5.59/lb (£12.31/kg) for 'good summer kid', the best quality finest fleece. Angora goats are sheared twice a year, the annual fleece weight produced by an adult female being 11–17½lb (5–8kg). The yield from a kid is of course much less, about 3¼lb (1.5kg) per annum.

Fixed costs per goat:
These are similar to costs for the dairy-goat, but with less sophisticated buildings, depreciation is less: total £45.

Variable costs per goat in 100-goat herd:
These are less than for the dairy-goat, assuming that grazing land is available. Feed costs are less because milk is the most expensive item for the body to produce, in terms of nutrition, and the Angora goat produces only sufficient milk to rear her kids:

Concentrates	£15	
Forage	8	
Straw	2	(housed for kidding, etc.)
Veterinary	8	
Shearing	3	
Transport	2	
	£38	

Financial return per female Angora goat:

Adult clip (i.e. sale of fleece)	£12.5
Kid clip	11.0
Sale of kids	90.0
Sale of culls	3.0
Less cost of rearing replacements	−2.0
	£114.5

Profit per goat:
This return of £114.5 – (38 plus 45) is £21.5 (figures prepared by Goat Advisory Bureau). It is noticeable that this profit depends on the goats' obtaining almost all of their nourishment from grazing/browsing on the farm, and on the selling of kids for £90 a year. The profit margin could therefore be very vulnerable; on the other hand an improvement to the yield and/or the price of mohair would increase profits considerably.

It was stressed that with farming dairy-goats, finding a market is half the battle. Mohair producers who choose to join the co-operative, British Mohair Marketing (BMM) can sell their fleece output through BMM, so marketing is not a problem. However, better prices can be obtained if the fleece is processed – spun and knitted into designer garments, for instance – but then markets must once again be found. Hand-spinning is time-consuming. Some of the time can be saved by having the fleece carded, though costs run at about £3.18/lb (£7/kg) for a minimum quantity of 11lb (5kg). Machine spinning services are also available, priced at around £7.27/lb (£16/kg), minimum 44lb (20kg).

The feasibility study stressed the financial benefits of selling surplus animals for meat. The British Angora Goat Society have made great efforts to promote goat-meat to increase the demand and enhance the price. The trade mark 'Capra' was mentioned in Chapter 1, but there is a long way to go before goat-meat is in wide demand throughout the UK at a good price.

Cashmere goat farming

Cashmere and mohair are as different as chalk and cheese (see Chapter 14). The bulk of the world's cashmere production is processed in Scotland into expensive garments. The best cashmere, produced in Mongolia and sold through China, forms most of that bought by Scottish processors. Difficulties in this trade have created an interest and financial support for the development of home-produced cashmere. Therefore marketing your annual fleece crop is no problem, at prices around £36.36/lb (£80/kg). The trick is to breed goats which produce a reasonable quantity of the very precisely defined down required!

While Angora goats come from warm, dry environments, cashmere-producing breeds are those which need to keep warm in the worst winter weather. The cherished under-down grows in response to shorter day-length and, having served its purpose, moults out in the spring when it can be collected by combing (the Mongolian method) or shearing. The goats are thus hardy and, though shelter from rain is always welcomed, they can spend much of their time outside, except for kidding and, if they are sheared, after shearing.

It is not necessary, however, to subject genetically productive animals to arctic winters for them to grow fleece. Most, though not all, cashmere farms are in the Borders and Scotland, where the goats' function in improving hill grazing is appreciated. The Scottish Cashmere Producers' Association bi-monthly newsletter gives information on all relevant matters, including the progress of the meat-marketing co-operative for goat-meat produced in Scotland. The price paid for meat, however, may barely cover the rearing costs, and no more than 10½oz (300g) per year of fleece can be expected per goat at the present state of breeding-up. While a gross margin of £40 per female has been suggested, this could be wiped out by capital costs. It has been stressed by those engaged in it that cashmere farming is a long-term (10-year?) project, not one for those looking for a quick return.

Meat production

Many consumers would like to buy kid-meat, and a great number of goat owners would like to sell it to them. Market structures bringing the two sides together are unfortunately far too few in number. The Boer goat, specifically developed for meat, is still something of a rarity, while surplus kids and older animals from milk and fibre producing enterprises are numerous. Unless animals are sold live for meat, which cannot usually be commended on welfare grounds, and in any case is unlikely to pay even for rearing costs, the Food Safety Act applies, with its requirements for hygiene, refrigeration and registration of premises.

Slaughter to produce meat for sale for human consumption can take place only in licensed slaughter-houses, where charges run at £5–£10 per animal, higher than for lambs due to the increased difficulty of skinning goats. Some slaughter-houses will sell the skin back to you for about £1; these can then be cured for around £12 each and used or sold as rugs or for other purposes.

Rearing costs vary tremendously depending on whether kids are reared on their dams, or on goats' milk which would otherwise be thrown out, or on milk replacer. It can well cost £40 to rear a kid to eight weeks on bought-in milk powder. Dairy-breed kids are large enough to slaughter at five or six months, but the smaller fibre breeds need to grow for about a year, by which time a price of 90p/lb (£2/kg) will not do much more than cover costs.

To sum up, there are goat farmers making a living, but many have gone to the wall. It is a particular personality who can make a success of entirely market-led, unsubsidized farming, but those who do so evidently find it extremely satisfying.

CHAPTER FOUR

The Control of Goats

Much popular prejudice against goats, and many real problems of goat farming, arise out of the special difficulty involved in keeping goats under control. In cultivated areas, the goat's contempt for the normal stock fences, and its destructiveness to trees, gardens and crops are notorious; on waste land – mountain, heath and cliff – the tendency of the goat to return to the wild (turn feral) is a significant nuisance. Even when narrowly confined to a yard or loose-box, the goat contrives to trespass on the privacy and liberty of her owner by refusing to thrive or be silent out of her owner's company. For all these sins the penalty and solution may be found in the collar and tethering chain – the most laborious method of controlling farm stock, and one to which the goat is naturally ill adjusted.

It must be admitted that the goat is inherently less easy to control than other members of the farmyard community; yet the nature of the goat is disciplined, co-operative and intelligent; most of the difficulty in controlling her arises out of the fact that the goat's psychology, her requirements in the way of food and shelter, and the specifications for the fence that will control her, are so very different from those of other farm stock.

Psychologically, the goat has some particularly striking peculiarities. It is almost impossible to drive goats; they do not share with the other grazing beasts the convenient habit of packing and turning their tail to the approach of danger, dogs and omnipotent man; nor do they share with cattle and horses the instinct to put as much distance as time will permit between themselves and an object of suspicion. Goats scatter and face the enemy that comes suddenly upon them, committing their safety, not to speed and distance, but to superior agility and manoeuvre. The pursuit of a frightened cow or horse follows a bee-line, which can be directed by outflanking; the pursuit of a goat follows the course that the ragged rascal ran – round and round the rugged rock.

It is certainly possible to perform the illusion of driving a flock of goats, but only when there is a personal and private arrangement between the goatherd and the flock. In reality the flock is being led from behind, as the king billy of the wild flock may often lead; if the goatherd who is 'driving' the flock turns about and goes home, so does the flock.

The unfortunate Mary was able to call the cattle home across the sands of Dee, so long as her call meant relief to the cow with a distended udder or food to the cow with an empty belly. To cows with full bellies and empty udders Mary calls in vain, and has to send the dog to fetch them. But a goatherd can call his flock of hungry goats away from their foraging, and they will follow him, complaining but obedient, to be shut up in a cold yard for the night.

Yet should Mary let her cattle roam the sands of Dee with their suckled calves from year's end to year's end, and never set eye on them, she and the dog can still drive them home when required with little more trouble than usual. Let goats roam unattended and unrestricted with their suckling kids for a few weeks, and, if you see them again, they will be wild creatures; there is a slim chance that if you follow the flock slowly and call to them in familiar tones they may eventually respond; but as for using a dog on them, you are as well advised to send him after the red deer.

The psychological relationship between man and his domesticated animals, horses, cows, sheep, dogs, and poultry, is specialized. The behaviour of these animals towards man is different from their behaviour towards any other creature, and different from the behaviour towards man of their wild counterparts in captivity. The relationship is the product of centuries of selective breeding, and may be said to constitute their domestication. In this sense goats are not domesticated. At times they treat man as they treat members of their own species, at other times they treat man as their wild ancestors do in a natural state. There is no qualitative difference between the behaviour towards man of wild goats captured young and that of so-called domesticated goats.

Anyone who has the care of goats soon grows to realize that the relationship between the goatherd and his flock is a great deal more personal, more intimate and more delicate than is usual in the farmyard. It is in fact similar to the relationship with a gregarious animal that has been tamed from the wild. But those who tame and control wild animals usually have an advantage that goatherds often lack – a good knowledge of the social behaviour of the animal they are taming. It is a great deal easier to control goats if you understand something of the social structure and routine of the wild flock on which the behaviour of domesticated goats is based.

The wild goat flock consists of up to thirty or forty goats of all ages and both sexes. In structure it is an easygoing patriarchy. The king billy of the flock rules by right of strength and courage; he maintains discipline, keeps the peace and coherence of the flock and is its constant guardian. He leads the single file in peaceful movement, but shares the practical leadership of the foraging expedition with an old she-goat, the flock queen, who will normally outlast a succession of kings. Billies are expendable; their great size, ever-growing horn and hair and extravagant sexual habits throw a constant strain on them; their fearless readiness in defence of the flock secures for them an early and heroic end if the opportunity presents itself. Yet the billy is utterly egocentric: the flock is for him an extension of his own person; his care for the flock is but arrogance, and the flock but render him his due. They accord obedience while he is present; but they do not lament his absence. It is the old she-goat, the flock queen, who is the mainspring of the life of the flock.

When she stops to feed, the flock feeds, when she raises her head from browsing and stares at the billy, the king moves on to the next foraging site. In the face of danger she leads the flock to safety, while the billy brings up the rear and holds the enemy at bay. She sounds the alarm if any member of the flock is missing, and will not be content until he is found. When she is absent the flock is in turmoil. If the flock grows too large for its range, it is an old she-goat that leads the breakaway party, with a young billy as guardian, to form a new flock.

The flock is coherent throughout the greater part of the year. Most of the kids are born within the space of a few days, and at this time the goats in kid break away from the flock for about a week, the king billy keeping the remnant together till the flock re-forms. For some weeks thereafter the flock queen is too much occupied with maternal duties to take much part in flock leadership, which rests almost entirely on the shoulders of the billy; at this season, when his protection is most needed, he enjoys more power than at any other.

About a month before the onset of the breeding season, the king billy is much occupied in chastising his sons and grandsons, the mothers cease to call their kids and the flock travels long distances on to poorer grazings in a very straggly formation. Only at nightfall does the flock re-form during this period; it re-forms where its wanderings find it, and does not return to its citadel for the night as it does at other seasons. This late-summer wandering serves to wean the kids and alter the flock's diet from the soft, milk-producing forage of summer to the hard, heat-producing forage of winter.

When man enters the social circle of the goat flock, he (or she) assumes the rank of kid, flock queen or king billy, according to circumstances. As kid, man can milk the flock peaceably; as king billy or flock queen, man can lead the flock. If the goatherd has nothing to do but milk and lead his flock, he will have no more difficulty in controlling his goats than the little boys, of seven to twelve years of age, who keep their flocks from trespassing in the surrounding fenceless fields all the way from Lisbon to Pekin. There is no great difficulty in getting into the social circle of the goat's confidence; the difficulty arises in getting in and out at your convenience and still exercising control.

Many goatkeepers start with the indiscretion of weaning their kids on to a feeding-bottle at birth, and keeping them separate from their emotional mothers until the fount of maternal milk and affection has dried up. The mothers so treated expend part of their energies, properly devoted to making milk, in calling for their human kid-substitute; and devise whatever means and mischief their ingenuity can contrive to bring their kid-substitute hurtling out of the house to chase and handle them. It is more peaceful and profitable to allow kids that are to be kept to suckle their mother for the first four or five days, then wean the kids on the bottle, but leave them in small kid-boxes, beside their mother in the goat shed, where they can be seen and smelt and heard, but not suckled, for a further ten days. Thereafter they can run with their mother without fear of their sucking her. Kids make good progress if allowed to suck their mothers on range. Contrary to popular opinion, kids so reared are more manageable when adult than kids reared away from the mother; emotional deprivation in youth does not make for emotional stability in adult life in man or beast; the goat has a deeper fund of patience, time and topical knowledge with which to educate and discipline her own kid, than any human mother-substitute.

The policy of allowing the natural social organization of the goat flock scope to develop is, in general, a sound one. The flock of intensively kept goats devotes itself more wholly and happily to milk production if it is allowed a modicum of communal life; yarding goats is preferable to stall-feeding for this reason among others (always assuming that the yarded goats are all dehorned or hornless). Unless the goats have opportunity to give a practical demonstration of their strength, agility and character to one another, they lack the data on which to appoint a flock queen. If there is no flock queen in the flock, her human attendant becomes for each member of the flock an indispensable companion in whose absence she feels insecure and ill-content. As goats are highly intelligent, they soon learn that the more noisy, destructive, wasteful and faddy with their food they are, the longer they enjoy the desired company.

To keep one goat alone, unless you are prepared to spend most of your life with her, is not only troublesome but extremely cruel.

If the goat is permitted to love its own kid, and the flock to embody their sense of security in one of their own number, it becomes possible for the goatherd to enter and withdraw from the flock without disturbing the even tenor of its way. The goatherd then fills the role of kid only briefly at milking times, and for the most part occupies the position of king billy in the social structure of the flock.

Even so, withdrawal from the flock is not always easy. It is easy for the goatherd to leave the goat shed or field and shut the door or gate; provided he leaves kids and a flock queen behind him, he does not leave a bleating vacuum. It is less easy for the goatherd to lead the flock out on range and to leave them there.

The king billy leads his flock in peaceful movement, with the flock queen close behind or beside him. Both know the rounds of their range in detail; according to wind and weather and season they follow a predestined course. King billy stops at an ash thicket; the flock queen starts stripping the young leaves; the rest of the flock follow suit; though other palatable forage be close at hand, ash leaves are the *table d'hôte* for all. King billy reaches highest and eats ravenously; flock queen bends down the branches for the smaller fry, but doesn't forget her own needs. After about ten minutes, the flock queen stops eating, approaches king billy, who is still stuffing himself, stares at him pointedly and, if necessary, makes a small remark. King billy stops eating and moves on. The rest of the flock who have scattered through the thicket, hurry to re-form the single file and follow. Next stop is for a patch of scrub oak (to balance laxative ash), where the same procedure is repeated. After perhaps a dozen stops and starts, the flock settles for the day on a suitable area of mixed browsing and there each one eats to repletion, lies down to cud and eats again. Towards evening king billy leads the flock back to their 'citadel' or goat shed by stages. Occasionally the flock queen or another senior member of the flock may stop for a moment to satisfy some personal requirement – say conifer bark – king billy waiting politely or holding on his way according to the importance of the goat who has stopped; but the rest of the flock does not feed.

The goatherd who knows his flock and range can lead on the first stage or two of their wanderings, and as soon as they are all busy at a popular stop he can quietly slip away. The flock queen will call him when she is ready to move on, but if he has disappeared she will take the lead herself. This tactic works well in broken wooded country; but in open country the delinquent leader is caught in the act of escape and the whole flock comes scampering after him.

Under such conditions the goatherd must use plain language to tell his flock when he has ceased to be their king billy. Mere rudeness won't suffice; king billy's manners are not of the best; a push and a grunt and a show of displeasure convey his warning to the flock to obey and follow him more closely. But king billy doesn't throw sticks or stones; the performance of such an act is a special characteristic of man and monkey which is peculiarly repulsive to all other species. So long as it is carried out calmly and ceremonially, so as not to be mistaken for mere rudeness, the flock will turn their backs on the goatherd, will go their way and leave him to go his. The popular use of effeminate males has produced flocks of half-witted goats which may be incapable of finding among themselves a flock queen of sufficient independence of character to accept even so broad a thing as the thrown stick. If such a flock refuses to be content without a king billy to lead them, the only cure is to give them a real one.

But the introduction of a male goat into the flock on range brings into close perspective a problem that is inherent in the relationship between the flock and the goatherd. Because the only way of controlling the flock is to be accepted as a member of it, the goatherd's position is always open to the same challenge as that which faces the flock queen and the king billy. Occasionally the flock queen may challenge his right to lead; an adult male with the flock on free range will always make the attempt at least once. Honeyed words and bribery are wasted on a rebellious goat. A good sound trouncing is needed; not just an angry slap, but a stand-up fight and a walloping thorough defeat.

As a prefect at school I was told to inquire into the circumstances whereby a certain small boy, of peculiarly retiring character, came to get his Sunday-best suit covered with long white hairs and peculiar odour each Sabbath day. I followed the small boy when he took the liberty of a Sunday afternoon walk by himself. He led me several miles across country to a small paddock in which there was tethered a very large white billy goat with awesome horns and ferocious habits. The billy must have weighed twice as much as the small boy, who approached him purposefully. Having taunted the billy to the end of his tether the boy grabbed him simultaneously by one ear and the tail, drew the two extremities together, forcing the billy to stagger round in a circle, downed him, rubbed his nose in the dirt, and then let him go. Then he did it all again. Satisfied, he returned to school and his retiring habits.

For the treatment of repression in small boys, better techniques may have been developed – but in seventeen years of goat husbandry I have been unable to discover a technique for the treatment of rebellious goats which is as effective, nor one which calls for less strength in the operator, nor one which is less painful for the goat. The need for some such treatment is widely acknowledged; for lack of a humane and practical one, recourse is had to hunting crops and other inhumane and impractical measures.

Not that the male goat is normally obstreperous – on the contrary, a vicious billy is a rarity, and his ill nature due either to being tethered and teased by small boys and dogs, or to being closely confined and handled with fear and distaste by attendants unable to control him.

In the wild flock the king billy is one of the most tolerant patriarchs of the animal world: he puts up with immature and adult males; personally instructs the youngsters in good manners and flock defence; and only hunts them off during the height of the breeding season – and then with more formality than effect. Young males happily accept the lead of the king billy, and are equally prepared to accept the goatherd as his substitute. But an adult male, though normally amenable will, if run with a flock on free range, challenge the goatherd's lead at two seasons of the year.

During the period which follows the birth of the kids, when the flock queen is occupied with maternal responsibilities and the billy's defensive potentialities enhance his importance in the eyes of the flock, an assertive male will often try to take control out of the goatherd's hands and to turn the flock feral. But the call of the milking pail has so urgent an appeal to the freshened goats, that, though the goatherd may have to go and fetch the flock for milking, there is little fear of losing control when he finds them.

But in the month before the onset of the breeding season, there occurs an unfortunate community of purpose between the self-assertive male and the milker who is trying to wean from the milking pail. In wild country there is a very serious danger of the flock turning feral, whether the billy runs with the flock or not; the danger is greater if he is with the flock.

The control of goats is mainly a matter of preventing them from trespassing, and in wilder districts from turning feral. The control the goatherd can establish is effective so long as the goatherd is present, and, in so far as he can inculcate a routine, it is effective when he is not present. For a free-range herd, the ideal is to establish a series of circular tours, each appropriate to a particular season and weather, and none of them making direct contact with forbidden ground. The initial training takes much herding time, but once the circular tours are established, the flock will not place a hoof outside its territory unless the goatherd leads.

For nine years we herded up to thirty-two goats on free range over 1,000 acres (405ha); some 300yd (274m) from the goat sheds a colourful garden lay unprotected and easy of access under their eyes each day. Only once was it raided – by a 'new boy' in the flock; it just happened to be off their beat. So did a patch of brambles which were a death-trap to sheep; a month of regular herding was needed before the goats would include it in their self-conducted tours. On the other hand, a wartime goatkeeper in Kent kept his goats in a large paddock separated from his garden by a thoroughly goat-proof fence. During a daylight air raid a plane dropped a stick of bombs across the holding: one in the garden, another on the fence between garden and paddock; while the whine of the departing plane was still in his ears, the owner poked his head out of his shelter to see what had become of the goats grazing in the paddock. They were eating cabbages in the garden. The garden was on their beat – despite the fence.

The minor problem of preventing goats from poisoning themselves arises only when the basic principles of goat control are ignored. Discrimination in matters of diet is a dim instinct in goats; education on the subject is a main feature of their social life. If the social life is stunted, by lack of communal activity and separation of mother and kid, not only are the members of the flock particularly liable to poisoning, but they have a jittery approach to all food, wholesome or not.

Safety from poisoning in the natural flock is achieved by the discipline of the foraging expedition, which demands that every goat eats the same kind of herbage, in the same patch and at the same time as the leader of the flock. Though most of the older goats may liberalize this discipline to some extent, it provides a constant fount of instruction as to what may be eaten, and where and when. In a flock with the active communal life, it is amusing to test their reactions to unfamiliar food, by leaving some unfamiliar but acceptable article, such as red cabbage, in their way. You will be rewarded by a grand display of ham acting.

The first goat to spot an unfamiliar food stops as if shot, snorts, and adopts the tense stance with feet set wide and ears pricked, exactly like a red deer hind who has caught a whiff of stalker. Then comes a cautious, inquisitive approach, nose stretched forward and the body bunched for lightning retreat. Then the flock queen takes over the investigation, adopting the same poses, and the rest of the flock huddle and watch. First the flock queen noses the object; then she nibbles a bit and spits it out; nibbles again, and chews tentatively; if good, she swallows it, and, as if revelation had suddenly come upon her, proceeds to wolf it. The other members of the flock, after a cautious sniff of identification, follow her example; each secures a morsel at least. If the mysterious food is not good, the flock queen spits it out with vehemence, snorts, fusses, runs frantically about wiping her mouth on the grass and pawing at it, and generally puts on a superb act of frightened disgust; then, having done justice to her station, she reverts to her normal foraging with sudden equanimity. Each member of the flock goes through the pantomime – the billy snickering his protective concern. When the others have finished their pantomime, the billy engages the offending herb in heroic combat, and may take half an hour in obliterating its presence: a memorable performance which must imbue the flock with a great sense of digestive security.

The only aid the goatherd can lend to the natural discrimination of the flock is himself to draw attention to, spurn and destroy any poisonous weed which evades the notice of the foraging flock.

Goats generally have good road sense: a little on the panicky side. Here, their reluctance to run directly away from an approaching menace is serviceable; they leap to the bank or the road fence and never run on in front of a car as sheep and cattle will. But a youngster may miscalculate the speed of approaching vehicles and try to cross the road to join companions on the other side. So all the goats should be kept on one side of the road; if tethered on the roadside, the tether's length should not permit them to reach the road. As goats grazing roadsides are not bona-fide road users (going from one place to another) they have no legal right there, and there is no redress for accidents, in law.

If these comments on the psychological aspects of goat control have been mainly negative – in the sense of preventing the goats from doing damage to themselves or others – it is because the positive job of maintaining a contented and productive flock is very much the same for goats as for other stock. Regular, peaceful routine is, perhaps, even more necessary to the naturally disciplined goat than to their more anarchic farmyard companions.

The normal routine, which the goatherd aims to inculcate, is for the goats to forage purposefully and peacefully from their feeding racks, fields or free range, to cud in comfort and to come to the milking willingly and without excitement. If they keep out of mischief in the process, so much the better.

However idyllic the relationship between the goatherd and his flock, this pattern is liable to rupture if the goats are physically uncomfortable. The commonest form of discomfort is occasioned by the weather.

Here we come upon a curious anomaly. Goats are notorious for their objection to wind and rain, and for the haste with which they rush to shelter from the least inclemency; and goats are famed for their hardy endurance of the windiest and wettest regions of our climate on the mountains and cliffs of the Atlantic coast. The feral goats of the mountain tops are but a generation or two removed from the hot-house pets of the goat house. Many a winter I have seen our milk-recorded herd of British Saanens grazing a ridge in the face of a bitter north-east sleet shower which has driven the thick-coated hill cattle to shelter, and sent even the mountain sheep to a warm bed in the heather.

If the goatherd wants to evoke from his flock the hardiness which is inherent in all goats, first he must consider the profound difference between the central-heating systems of goats and men. Man derives his heat almost entirely from the oxidation, 'burning', of the carbohydrate reserves of his own body. In cold weather he eats foods rich in digestible carbohydrates (in the fuel to bank his fires), and he insulates himself in warm clothes and houses. The main source of heat for the ruminant is that produced by the bacterial fermentation of fibre in the rumen. The fuel which supplies most of the heat has not been digested by the animal; indeed, it is material that no animal can digest. So the ruminant meets the challenge of cold weather by eating yet more indigestible roughage, and so lowering its intake of digestible carbohydrates.

The central-heating system of all ruminants is the same in principle, but the goat carries the principle to its extreme. The goat, which can in a day eat twice to three times as much as a sheep of the same size, is naturally designed to exploit pastures too poor to support the sheep; to live on them the goat has to eat to capacity. Consequently it is the fate of the goat to carry about with it for ever a radiant heater with two or three times the heat output of that of the sheep, and to carry that sweltering load both winter and summer. Mercifully, she has a thin skin and only a light covering of hair in summer, which is reinforced by the growth of underfluff in winter. Even so, you may examine the necessary generosity of her sweat ducts in the grain of morocco leather.

Being designed along lines which should provide a constant surplus of heat, the goat is sensitive to chills. The temperature at which her digestive processes work best is a relatively high one, and cold conditions of the rumen favour the activity of agents of disease.

Under domesticated conditions, especially when put out to graze good-quality pastures, the goat is often offered a diet so poor in the fibre that is her main source of heat, that her rumen temperature is dangerously low and even a summer shower may chill her perilously. She comes into her shed in the evening so puffed up with cold mush that, unless her udder be laid on an insulated floor, she gets chilled again. To protect her from these risks, her owner builds a fine, cosy shed, which warms up quickly when the goat comes in with a little autumn roughage in her. When she goes out the following morning, with nothing but a cold, wet pudding of concentrates and mashed turnips in her, she feels the cold more intensely. The nice, cosy shed will also ensure that the underfluff in the goat's winter coat never develops fully (some scrupulous goatkeepers have been known to comb it out too); so the winds of winter strike on her bare skin. When a really cold spell arrives, the goatherd puts a suet pudding on the stove for herself and an extra pound of maize in the goat's feeding bucket; so the goat eats even less roughage than usual and ungratefully – but logically – dies of pneumonia.

If it is desired to feed goats a highly concentrated diet, as it well may be where land and labour are scarce, then the goats must be of a suitable type, the housing must be consistently warm and the goats must not be expected to supplement their diet by foraging outside in any but the mildest weather.

When goats are expected to obtain a good proportion of their food by foraging for themselves, it is essential that their shelter never becomes heated much above the temperature of the outside air. In cold weather, especially, it is bad practice to feed concentrates or succulents (roots, kale, etc.) unless the goat has an adequate quantity of fibre inside her. The state of affairs cannot always be tested by eye, as the goat may bulge with kids, gas or water; but it can always be tested by pressing with one's knuckles in the left flank: where fibre feels 'puddingy', kids are absent, gas exerts counter-pressure, and water sloshes. In summer-time an alternative to highly nutritious grass should be provided in the form of hay, if nothing fresh and natural is available in the way of roughage. This will improve the goat's appetite for grass, and enable her to graze in cooler and wetter weather.

In winter, dry goats require very little in the way of nutrients, and can get all they need in an hour or two on heather and a little longer on brushwood, tree bark and brambles; there is no point in keeping them out of their shelter when they want to return to it. But they are excellent weather prophets, and will always make the most of a spell of good weather if free to come and go as they want.

When goats have to be supervised the whole time they are grazing, it is most irritating if they refuse to get on with the job, and alternate between short spells of foraging and long spells of cudding. If they are led back to their shed as soon as they start cudding, they will soon learn to take as much as they can hold in one whack, and so save herding time. Though food consumption under this arrangement is increased, the goats do not make such good use of the food they eat, either in the field or for several hours thereafter. The goat's digestion works best when relatively small quantities of fodder are passing rather rapidly through the rumen; the more nutritious the herbage, the greater the losses if fed in big feeds. It is therefore better to herd the goats out for half an hour in the morning and half an hour in the evening than to herd them for two hours on end. This is lighter on labour and feeding and the goat's digestion, and makes a heavier job for the milker.

Under most conditions, unfortunately, the control of goats requires more than a good command of their mental and physical comfort – it needs fences.

Any fence less than 4ft (1·25m) high will be jumped, but over 3ft 6in (1·1m) high it will not be cleared by the milkers, who will bruise and leave portions of their udders on the top of the fence. The only exception to this rule is the electric fence, which goats are understandably loath to jump even at 3ft 6in. A standard 4-ft stock fence, with seven wires spaced at 6in (15cm) intervals, will hold goats of a reasonably placid disposition so long as the wires remain dead tight. Eventually the goats will force their way through the wires. At any time a sudden urgent incentive will send most goats over the top of a 4ft fence of this type. In theory it is possible to keep wires taut, but the theory ignores the fact that wire expands and contracts with changes in air temperature; if tight on a hot May afternoon the wires will probably snap in the frost of the following morning, unless they are 6-gauge or heavier.

Wire-netting fences give a great deal of pleasure to goats, who derive exquisite satisfaction from massaging themselves on the netting. The procedure is for the flock queen to throw her full weight against a section of fence between two posts, and to press her flank slowly along the condemned section; each member of the flock follows her, and then she will take another turn; the game goes on for hours every day, and for as many days as the netting lasts. When the section has collapsed and the goats have had their fill of the crops on the other side, they lie down to cud on the ruins, which provide that degree of insulation from the earth which goats crave. The hard, ugly glint that appears in the eyes of some farmers when goats are mentioned is often due to the pleasure that wire netting provides.

If you run two lines of barb along the wire netting it spoils the fun, but usually does some serious damage sooner or later.

With fences of the Rylock and Bulwark type goats adopt a rather similar technique: they provide massage in the first instance; but with fences of this type it is sufficient that the fence be given a slight slant, at which stage the goats accept the convenience of the ladder steps provided by the horizontal wires, and go over the top. There is then the added embarrassment that they cannot get back again to their own ground when occasion demands.

The most generally applicable solution to the fencing problem is a 4ft (1·25m) fence of chain link, with posts at 6ft (1·8m) centres, and with two or three support wires. The chain link can be either 2in (5cm) mesh and 12½-gauge or 3in (8cm) mesh and 10½-gauge. It must be strained.

The method of erecting this fence is as follows: after lining up and erecting the fencing posts and putting the support wires up at 4in (10cm) and 4ft 4in (1·35m) above average ground level, link up the 25yd (23m) sections, in which the fencing is sold, for the required distance, and lay the fencing flat down beside the fence. If the fence passes over any large humps or hollows the wires of the chain link are unlocked over the appropriate sections at both top and bottom and twisted up or down to give the required curve to the bottom line of the fence. Then they are relocked. Now fix one end of the fencing to one corner post and staple a 4ft (1·25m) length of 3 × 2in (8 × 5cm) timber across the other end of the fencing at a point which is about 3yd (2·75m) plus 1yd (0·9m) for every 25yd of fence length, from the other corner post. Then take a straining wire from the dead centre of this piece of timber to the corner post, taking care that the wire doesn't pass through the fencing. Finally strain this wire, feeding the end round the corner post until the fencing is rigid. Make fast the straining wire, and staple the loose end of the last 3yd of the fencing to the corner post; should the fencing subsequently slacken, straining on the one wire only will tighten it once more.

An electric fence is satisfactory for temporary fencing, but requires to have three live wires at 12in, 27in and 40in (30, 69 and 102cm) above ground level; these heights may need adjustment for large or small goats. The goats must be made acquainted with the potentialities of the fence, and learn to respect it, before the fence is charged with the duty of holding them. If they are allowed to break through once, they will accept the sting as the necessary price of liberty and do it again. Male goats sometimes appear to be entirely devoid of a sense of pain, and their control can never be entrusted to an electric fence.

In more permanent forms of fencing, where a solid top bar is a feature of the design, the height should be increased to 4ft 6in (1·4m) a height which is suitable for all gates into goat enclosures.

In pale or rail fences, the distance between the pales or rails should be under 3in (8cm) or between 7 and 8in (18 and 20cm) to prevent a goat or kid trapping its head in the space. The wider spacing is not kid-proof. The same limits apply to the distance between rigid bottom rails and ground level.

No type of hedge associates well with goats. If a hedge exists an electric fence should be placed well inside it so that the goats cannot reach the hedge. A wire mesh fence just inside a hedge will be used as a useful foothold to enable the shrubs to be consumed. If possible,

Plate 4 An electric fence is all that stands between Toggenburg temperament and a field of kale. Note the heights of the three wires.

seriously poisonous shrubs should be dug out, or cut down regularly, otherwise one day there will be a tragedy. A mature beech hedge of 5ft (1·5m) or more high will sustain little damage if alternative leafage is available; if grown in the Continental trellis style, such a hedge will be goat-proof as long at it lives; but the normal double-row beech hedge will usually admit a determined goat. Thorn and blackthorn hedges may hold goats as long as they live, and a pure blackthorn hedge will survive many years with goats; but they are rare, and the normal thorn hedge will not stand up to persistent goat browsing. A yew hedge is too dangerous to test. A holly hedge is a goat's paradise. Barberry is also eaten but not avidly. In general, hedges may control passing goat traffic but not persistent attack.

The effectiveness of stone walls as a barrier to goats depends principally on the roughness of the surface. The width of the top is of no significance. Nothing under 4ft (1·25m) sheer height is any barrier at all; beyond that height the goat cannot make a clean jump (billies to 5ft [1·5m]); but if the surface of the wall is rough, with a few projecting footholds, they may run diagonally up anything up to 6 or 7ft (1·8 or 2·1m). As with all goat control, much will depend on the social life of the flock (an elderly flock queen is not given to acrobatics) and on the degree of frustration they experience in their confinement.

While the use of barbed wire in goat fences is generally neither safe nor effective, the addition of one strand of barbed wire at 5ft (1·5m) on the 4ft (1·25m) chain-link fence makes a combined goat-and-deer-proof fence which is quicker to erect and requires less maintenance than a classic all-wire deer fence. Where goats are kept on a hill farm, such a fence has been held eligible for subsidy under the Hill Farming Acts.

In general, it can be said that a fence that is goat-proof is also deer-proof and vice versa. Few deer fences of the 11-wire type are deer-proof or goat-proof; but most of them are a sufficient deterrent for practical purposes.

Many a poor man cannot afford a fence against either deer or goats. The deer carries away a ·22 bullet in its guts, and the goat goes on a tether.

One goat alone on a tether is not a happy sight. A number of goats tethered within sight of each other can form a contented and productive community. Granted that the tethering system has been much abused and that it is very expensive in labour, it provides a perfectly feasible and satisfactory method of goat control under some circumstances. Much of the unqualified objection to tethering is rooted in plain, common snobbery. On the other hand, many goats have died, strangled, on the end of a tether. Kids must never, never be tethered.

The commonest form of tether is the picket tether, where the goat is tied by collar and chain to a stake driven into the ground. The collar is the better for being extra wide; the chain must have swivels at both ends, and be reasonably light. It is helpful if the chain is made of three easily detachable 6ft (1·8m) lengths, so that the tether can be lengthened as the day progresses, without excessive fumbling. The chain is attached to a ring which rides and revolves freely in a collar round the top of the stake. The stake must be driven in until the ring is flush with the ground, or the top of the stake will foul the chain. If the local blacksmith is too busy mending tractors, tethering equipment is available from the specialists in goat equipment who advertise in the appropriate papers. Pig tethers are adaptable.

The main snags of this form of tether are that they make for uneconomical use of ground, and can only be used on patches of grass and soft herbage. The goat grazes a ring at the end of its tether and soils the circle around the stake; if you divide a field into rings which do not overlap, you leave a lot of ground ungrazed. As a light chain will snag on any obstacle from a fruiting dock stem upwards, the ground on which it will not snag can be guaranteed devoid of the harder types of herbage in which goat digestion delights. In warm weather, the picket-tethered goat is often enveloped in a thin haze of flies which detract from its peace and content, and from which there is no escape. A spray with insect repellent (rather than an insecticide) helps matters in this respect.

The alternative form of tether, known as the 'running tether', is slightly more cumbrous to establish, but considerably more satisfactory in action. Essentially the running tether consists of a fairly short length of chain – not more than 3ft (0·9m) and not less than 18in (45cm) – with a swivel, collar and goat at one end, and at the other end a swivel ring which runs along a wire stretched taut along the ground.

The wire may be of any length from 6 to 7yd (5·5 or 6·4m) up to several hundred. If of a short length, the ends of the wire will be held by two stakes, the heads of which are driven flush with the ground. Alternatively, a long wire may be lightly strained from the fencing posts across the corner of a field, or right across a narrow field. In this case the goats' chains must be sufficiently short to prevent any possibility of their even considering jumping the fence at each end. A long wire can also be strained between two posts planted in mid-field; but in this case a piece of wood, longer than the length of the tethering chain and goat combined, must be tightly stapled to the wire at both ends to prevent the goat from getting caught up on the post. Such a piece of wood can also be stapled on to the wire at any point along its length to divide it into tethering sections. In this way a dozen or more goats may be tethered along the 'grazing face' of a forage crop; the daily move forward will only call for the removal of the end posts and one strand of wire; the cost of such equipment will be little more than the price of the wire. As against the electric fence, it has the disadvantage that you must spend a few moments attaching each goat to its tethering chain each day, and the advantage that there is no electric circuit to be shorted out by a heavy, wet crop.

The running tether uses a short length of relatively heavy chain which does not snag easily; if it does snag, the goat can apply a great deal more force to release herself than is possible on the picket tether.

The running tether has proved quite satisfactory for strip grazing of kale, with grazing periods of up to four hours. It has proved entirely satisfactory for goats eradicating thistles and docks from new leas for grazing periods of twelve or fourteen hours; it can be used with a measure of watchfulness right through the midst of a strong old growth of brambles and briars.

How much wind and rain the tethered goat can tolerate depends on the kind of fodder she is getting. It is quite proper for the owner of the goat receiving 4 to 5lb (1·8 to 2·25kg) of concentrates a day to rush her corn bin back to the goat shed as soon as a rain cloud darkens the face of the summer sun. But sterner diets make goats of sterner stuff. If an elementary shelter is strategically placed at the windward end of a running tether, most goats can be tethered on coarse weeds and grass most days from May to September.

Flies and sun are another problem. As far as the former are concerned, the goat on a running tether is little worse off than the goat with free range over a few low-lying acres. Sun intensity in Britain seldon reaches the goat's limits of endurance, provided the goat has access to water; but in damp, windless heat the goat is best kept in the shade during the worst of the day: it is not the sun that troubles the goat so much as hot humidity.

For any goatkeeper who is not a pedigree breeder of the first class, there is no purpose served in rearing male kids for anything except meat. There is always an ample supply of pedigree males at prices well

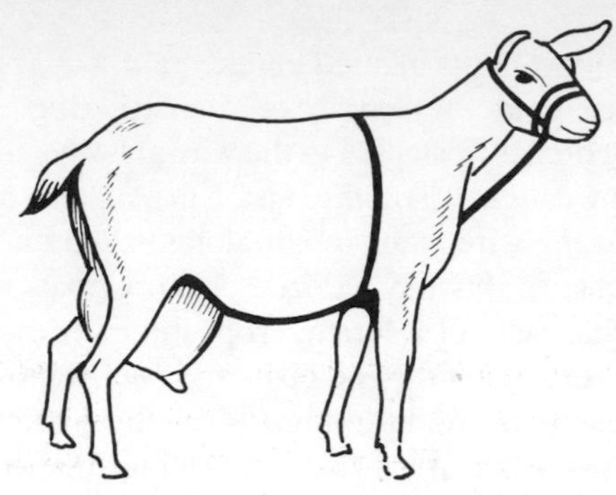

Fig. 3 The Spanish halter. A device used in both Spain and Germany on goats grazing mature orchards and olive groves, to prevent them from reaching up into the branches.

below the cost of rearing. Readers who care to disregard this excellent advice must accept the consequences and a very troublesome task of control.

From the age of three months, the male kid may be capable of service; outside the normal breeding season it is unlikely that he will give effective service, but the risk exists of spoiling one of the female kids with whom he is running, or all of them for that matter. For this reason males are usually segregated from the age of three months; the old males living in isolation are apt to maltreat male kids, so that each male kid has to be reared by himself and have all his food carried to him and all his social education provided for him by the goatherd. But some older males willingly accept a male kid for company. He must be taught to lead, and to obey, and not be permitted to play. Above all, he must not be teased.

The adult male is dangerous to all children and to any man or woman who is afraid of him: he is a nuisance to anyone whom he thinks might be afraid of him. To keep him in manageable condition, he must be handled and exercised regularly and very firmly. The method of reducing an unmanageable billy to subjection has already been detailed.

Fences to hold adult males must be at least 4ft 6in (1·4m) high; if with a solid top bar, they need to be 5ft (1·5m). Yarding, stalling equipment, gates, etc., must be of the strength appropriate to control cattle or ponies.

CHAPTER FIVE

Housing

Over a number of years I have been walking, through wind and rain, over the downs, over the hills of Northumberland and Galloway and among the mountains of Wales and Scotland. In all these places, and in all the wind and rain, I have stopped every hour or so to light my pipe; very seldom have I failed in the attempt. Wherever there is a wealth of ground feature and rock, there is a wealth of nooks and crannies wherein one can light a pipe in a wet gale. In such nooks and crannies the wild goats make their bed. The first duty of a goat house is to provide equivalent shelter; quite a number of goat houses fail to perform it.

The wild goat is a great deal more particular about its dormitory than the deer or hill cattle or hill sheep. Every dormitory of the wild goat that I have seen, in Galloway, Wales and the Scottish Highlands and islands, has boasted a carpet of a depth and insulating capacity unobtainable even at the Dorchester Hotel. Goats originated the principle of underfloor heating, of which the finest example I have seen was in an abandoned shepherd's cottage, near Back Hill o' the Bush in the Galloway hills, the winter quarters of the feral goats which summered around the Monument on the Newton Stewart road. Here the goats reclined on a 4ft (1·25m) deep compost of accumulated droppings.

The goats in the caves of Carradale Island had a bigger area to cover and had not achieved the same depth – but even the summer couches of the goats of Lochbuie in Mull were quite luxuriously lined. The need for luxury in this respect is clearly considerable; the need may explain why the feral goats return each evening to the same sleeping place, and the domesticated goat returns unbidden to its shed. When she has too soft and rich a fill of early summer herbage, or when snowstorms and torrential rain prevent her from getting an adequate bulk of winter roughage, it must be a real problem to the thin-skinned goat to keep her belly warm at night.

A further aid to the solution of this problem is found in the social nature of the feral goat dormitory. On warm summer nights, each goat has its separate couch; but if conditions are cold, there is much sharing of beds, and only one flank of each goat is exposed to the night air. Undoubtedly many cases of pneumonia, gastro-enteritis and mastitis might never develop if the goat were free to provide herself with a companionable hot-water bottle on the critical evening!

These features of the feral goat dormitory are not necessarily desirable in the goat house because they are natural. But there is little doubt that they are natural because they are desirable.

Our goat houses must inevitably be, to some extent, a compromise between that which is most comfortable and health-giving for the goat and that which is most convenient and economic from the point of view of human management. The design of cow houses has been inspired, most markedly, by the desire to ease the disposal of the dung the cow makes while she is being milked, fed and sheltered. But goat houses are designed for the prime object of enabling the feeding of concentrates.

Under most systems of management, it is economic to feed the goats concentrates at some time of the year. But the appetite of the goat is so elastic that one strong and hungry goat is quite capable of eating three or four times the ration prescribed to her, with ill consequences to her own health and the productivity of her companions. So goats can seldom be trusted to share their concentrates equitably at a trough, like cattle and sheep. They must be isolated from one another when consuming concentrates.

There is a remarkable unanimity about the layout of goat houses as a consequence. The goat house consists, essentially, of a feeding passage either side of which the goats are loose-housed behind feed barriers, or penned in individual loose-boxes, depending on the number of goats to be accommodated.

In this age of legislation, there are legal points to be considered during the planning of goat housing. Late in 1988, a new General Development Order came into force, the effect of which is that any proposed new animal housing within about ¼ mile (400m) of human habitation needs planning permission. So seek the advice of the local planning officer before starting.

There is welfare law to comply with too. The Welfare of Livestock Regulations 1990, which cover all goats involved in trading – whether it be the sale or bartering of products, livestock or manure – require them, when housed, to have adequate lighting, fixed or portable, to be available, sufficient to enable them to be inspected thoroughly at any time. The interior of any building, including the floor, to which the livestock have access must be constructed and maintained so that there are no sharp edges or protrusions likely to cause them injury or distress. No inadequately constructed or insecure fittings must be used for restraining the goats. When housed, they must be kept on, or have access at all times to, a lying area which is well drained or well maintained with dry bedding, and suitable accommodation must be available so as to enable any sick or injured livestock to be separated from the remainder, and this must have a suitably bedded floor. It is perhaps unlikely that goats would be kept in a house with an electrically operated ventilation system, but if they were, these regulations require there to be an alarm which would operate if the system were to fail in any way, and also a back-up ventilating system. The alarm and the back-up system have to be tested at least once in every seven days and any faults put right.

The Ministry of Agriculture's Code of Recommendations for the Welfare of Goats was published in 1989 (Ref. PB0081). To breach one of its provisions is not, in itself, an offence, but can be used in evidence to establish the guilt of anyone accused of causing suffering to livestock. The Code covers in some detail the adequate structure of housing and fittings (hay racks, etc.) for goats and should be consulted and followed.

This legislation applies to the housing for all breeds of goat. If, however, dairy-goats from which the milk is to be sold for human consumption are to be housed, there are additional requirements to be considered. The Ministry of Agriculture's Code of Practice, 'The Hygienic Production of Goats' Milk', based on an earlier Code published by the (then) DAFS (code number BL5677), has some recommendations on housing. This Code of Practice is likely to be similar to legislation controlling the production of goats' milk for sale for human consumption, expected to come into force on 1 January 1994. These recommendations are that goats should be housed in premises equipped with adequate lighting and ventilation and constructed and maintained in such a manner as to avoid undue soiling of the animals. All internal surfaces of the premises liable to soiling or infection should be capable of being easily cleaned. The floor should drain freely via a drainage system to a suitable place of disposal situated outside the building. Bedding should be maintained or changed so as to ensure that the goats are not contaminated by it. Suitable facilities should be provided for the storage and disposal of soiled bedding and droppings. Such facilities should not be situated in close proximity to the milking area or the milk room. Suitable and sufficient supplies of drinking water should be freely available.

When milk is to be produced for wholesale sale, i.e. for resale for processing or drinking, European law, to be implemented in the UK by 1 January 1994, requires that where goats are not kept in the open, the premises used must be designed, constructed, maintained and managed in such a way as to ensure good conditions of housing, hygiene, cleanliness and health of the animals. There must be isolation facilities for sick goats. There is provision for the regular inspection of the holding to ensure compliance.

The construction of milking parlours, also under legal control, will be dealt with in Chapter 11.

Individual pens (loose-boxes) should have a floor area of 2¼–3½ sq yd (1·8–3m²), depending on the size of the goats, for adult females; 1½–2¼ sq yd (1·2–2m²) for young stock, and a minimum of 4¾ sq yd (4m²) for males. Where goats are housed communally in large areas (loose-housing) an area of 2 sq yd (1·6m²) for each adult female should be

Plate 5 Inside the Malpas goat house where the modern dairy-goat has acquired several of her valuable qualities, one is impressed with the orderliness and careful detail that lie behind the successes bred there. A commercial dairy farmer would prefer fewer shelves on which coliform bacteria could accumulate in dung dust.

allowed; 1¼ sq yd (1m²) for each animal where youngstock are housed. Adult males will not be housed communally. It is vital in communal housing to allow enough feeding space for even the goat lowest in the 'pecking order' to feed. Trough or feed barrier lengths recommended are: 15¾in (400mm) for each adult female; 11¾in (300mm) for each young animal.

In comparison with the loose-box goat house, the communal goat house offers a high standard of liberty, of social amenity, interest and comfort at lower cost. This type of housing accords well with the methods of control and feeding advocated in this book, but is subject to one serious objection – the amount of bullying it permits.

Bullying does not normally reach serious proportions in a flock with an active social life and a recognized flock queen, unless some of the goats are horned and some hornless. If this is the case, either the horns or the communal goat house must be dispensed with. Also, as newcomers into a flock have to 'fight their way in', and the process takes some time in a large flock, the communal goat house is an unsuitable form where goats are regularly boarded, or there is much buying and selling of adult stock.

The first duty of the goat house is to shelter. Here we come to the standard instruction that it must provide plenty of fresh air but not draughts. The instruction is misleading, because the more fresh air there is in a house the less cause there is for draughts, and especially for the cold down-draughts, which are the most dangerous and discomforting. There is very little draught in the hollow on the hillside, but as soon as the warm air of a goat shed comes in contact with cold, thin-skinned roof and walls, a vicious down-draught is set up which strikes on the goats' backs; at the same time, cold air streams into the warm shed through every chink in the building and, being heavier than warm air, promptly seeks the floor where the goats are lying and swirls about them.

It is therefore very important to adapt existing buildings, or to build new ones, in such a way that good ventilation, without down-draughts, is assured. Professional advice on farm buildings can be obtained from the Farm & Rural Buildings Information Centre (see Appendix 3). The advantages of outdoor air over the air inside the goat house are not only the far lower burden of pathogenic bacteria and viruses, but also the much lower number of floating dust particles. These particles, when inhaled, overload the body's defence mechanisms, allowing bacterial infections to obtain a hold. Ventilation, then, is aimed at creating an atmosphere inside the goat house which resembles as closely as possible the purity of the outside air. Cold, within reason, is not a problem to goats over six weeks of age, providing they have a dry bed which successfully insulates them from the cold ground.

The warm stale air which rises up from the goats' bodies must be allowed to escape, and be replaced by air from outside, above the level of the goats. Many communal houses, which are always pleasant to walk into, have from a height of five feet (1·5m) up to the eaves, purpose-made plastic mesh, or Yorkshire boarding, which is vertical boarding with a space between each plank. This can be supplemented by an open ridge at the top of the roof, allowing stale warm air to escape. The walls and doors up to five feet from the ground are of solid construction (Fig. 4).

In very cold weather there could still be a problem with cold air falling from the inlets on to goats lying close to the walls. Hinged flaps can be fitted to deflect cold air when needed or, for small kids, stout wooden boxes on their sides will keep the cold air from kids inside the boxes. Kids will not be slow to take advantage of these cavelike sleeping places (Fig. 4).

Sick goats, and goats who have just produced kids, need separate pens, and this is recognized by law, as stated above. If permanent pens

Fig. 4 The ventilation of communal goat houses.

are not available, and a great many can be needed at kidding time, removable partitions, linked together by eyes and drop-rods, are a splendid way of making any stable or building into an instant goat house. If the partitions were designed for calves, however, it will be necessary to fix some suitable mesh over them to prevent the goats from simply stepping between the bars. Some form of fastening to the wall is necessary, and indeed the partitions come with wall-fixing plates; remember however that the walls to which they are fixed must be faced with some smooth impervious material for the sake of hygiene.

In severe weather, small kids must be kept sufficiently warm, as they do not have much ability to regulate their body temperatures. Infra-red lamps, suitably guarded and suspended above the kids' backs at the distance required by the manufacturer's instructions, may possibly be needed at very low temperatures. Table 1 shows the critical temperatures above which kids' pens must be kept at two and at ten weeks of age.

It may seem a little ambitious to recommend insulating a goat house. But in fact modern insulating materials are very cheap, readily available and easily handled. In most cases it will cost less to insulate a goat house with material to last its lifetime than to give it one coat of paint. Three main types of material are used: mica flakes, glass-fibre blanket and aluminium foil – all of them are fire-proof, rot-proof, non-absorbent and deterrent to vermin. Very poor meadow hay, loosely packed, is also a good insulator, but lacks the permanent virtue of the others. Mica flakes are laid loose, 1½in (4cm) deep, on flat or nearly flat surfaces. They are the cheapest and easiest form of insulation for flat concrete roofs, being covered with paper and sandwiched between two layers of concrete; they operate by trapping a quantity of still air in insulated pockets. Glass-fibre blanket works the same way, but can be draped over or under sloping roof surfaces. It is rather unpleasant to handle;

Table 1 Critical temperatures for kids (degrees C)

	2 weeks old	*10 weeks old*
Standing	10	−3
Lying on dry concrete	17	−6
Lying on dry straw	7	−8
Lying on damp straw	10	−3
Lying on wood slats	10	−3

either use leather gloves and close-woven overalls, or bare hands and arms and wash them in cold water when you are finished. Aluminium foil is not a blanket insulator but a heat reflector. The effect it produces is similar to that produced by the other insulators, but it is possibly better than any for retaining heat in a well-ventilated goat house, and for keeping a house cool in hot sunshine. It loses it properties if laid in direct contact with other metals, but is ideal for lining under felt roofs. It is tacked on the wall or roof surface in overlapping sheets.

Wood itself is a good insulator, and matchboarding is sometimes used for lining corrugated-iron sheds, but it is an expensive form of insulation.

The conversion of old stone- or brick-built buildings is usually the best means of providing warm shelter with a minimum of draughts. It is also usually the least hard on the pocket and the appearance of the countryside. But if new buildings are required, they need not necessarily be ugly. Bales of straw make good building bricks although they are no longer inexpensive; they can be secured in position between two lines of fencing. If set on a dry foundation, thatched on the outside and given a roof with a generous overhang, they will last for fifteen to twenty years. Thatchers can still be found at a price. Six tons of straw will provide walls and gable end for a twenty-goat house. Fire risk would be considerable, however.

In districts where straw is scarce, stones tend to be plentiful; for a warm, well-insulated shed there is much to be said for the primitive rough-stone house. It is not beyond the capacity of the average handyman to construct. The wall is made 3ft (90cm) wide at the bottom, tapering to 1ft 6in (45cm) wide at the top; the inside of the wall is kept straight and all the taper is on the outside. Cement is used on the inside of the wall, but on the outside only for the last foot at the top. Between the inner and outer stones at the bottom, turfs are rammed as building progresses. At a height of 3ft, strap stones are laid at intervals across the width of the wall; from that height upwards, the interval between inner and outer stones is filled with small stones and gravel. The outside corners are rounded, and the walls are taken to 6ft (1·8m) height or a little more all the way round – with no gable end. The roof, which must be angled or rounded at both ends, may be either felt or thatch. The width of this type of house must not exceed twice the height of the walls, and the roof must be rather steeply pitched. Good models of this type of house are common in the Hebrides.

Concrete-block construction is now relatively cheap, provided the

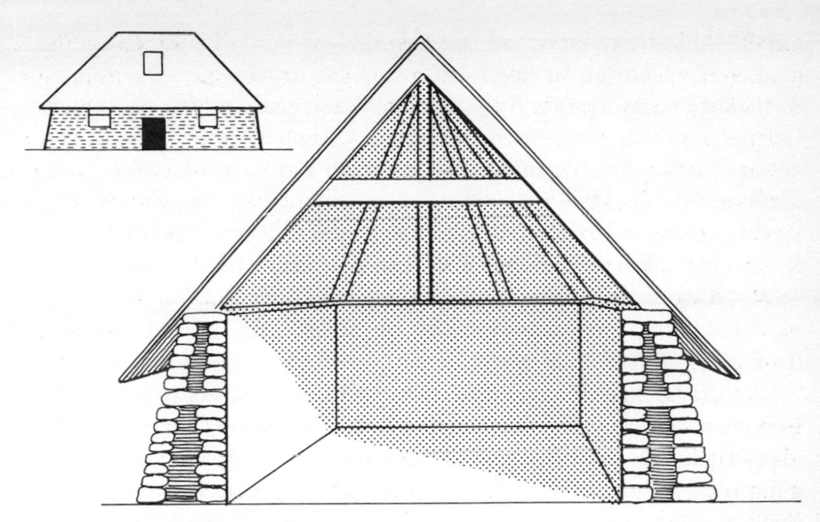

Fig. 5 The rough-stone house.

erection is not done by a builder. If there is no objection on the grounds of appearance, cavity blocks make a good goat house. Double-wall construction is preferable, but single wall will do. Flat roofs are the easiest and most satisfactory for concrete-block buildings, but they must be insulated. With flat roofs, two hopper-type windows or ventilators facing each other, opening inwards, with their tops at ceiling level, give the most satisfactory ventilation.

As male goats are large, strong and may want to escape to serve a nearby female, quite desperately, their housing needs to be extremely strong. Even a young and normally docile animal, which thinks a rival is serving a female goat, will apply himself to breaking out with terrifying strength; and it is natural that he should do so. The housing should be of concrete block or brickwork, doors and yard fencing of steel tubing. It is quite essential that the male goat should be able to be fed and watered without entering his house, i.e. buckets are suspended outside holes in the wall through which he puts his head to feed and drink. Male goats smell terrible in the rutting season and are therefore usually housed as far away as possible – but loneliness will not help his temperament, so he should be housed where he can watch plenty of activity, and see other goats.

The chill that strikes through an insulated concrete floor will reach through several inches of straw. If a new floor is being laid it is well worth while to insulate it by one or other of the following methods, neither of which call for much extra expense or labour.

Method 1 Lay the foundations of the drainage floor of heavy shingle or large chippings – no stone smaller than a hen's egg – unmixed, to a depth of about 6in (15cm) and form it into the rough design of your drainage floor. On this bed of loose stone pour a very sloppy mixture of 4 parts sand to 1 part cement. Allow 1¼–1½cwt (63·5–76·2kg) of cement for a ton (1,016kg) of stone. This will percolate through the bed of stones, lodging only where stones touch one another. It hardens quite quickly, and will give you a honeycomb of still air on which to finish your floor; have no fear of its strength, for the system has been successfully used on arterial roads. For a drainage floor it is preferable to finish the surface with a waterproofing compound. But Method 1 can be modified to produce a porous floor, in which case you will finish with a mixture of 1 part cement to 3 or 4 parts fine clinker and spread the surface with earth or sand.

Method 2 requires a tractor. Thoroughly till the building site to a depth of 9in (23cm). With a post-hole digger, excavate 6in (15cm) diameter post holes at 2ft 6in (75cm) intervals over the whole site. Fill these with concrete. Then as quickly as possible, and with the minimum of trampling, lay a first layer of concrete straight on the loosened earth. Finish your floor in the normal way, with a waterproof surface. In due course the earth subsides, leaving the floor supported on concrete pillars.

Such a thing as a standard goat house does not exist. The expenditure of architectural thought on goat houses has been small in quantity, if excellent in quality. It has therefore seemed more helpful to provide basic data and food for thought than specific designs for imaginary, and probably non-existent, conditions.

Good ready-made goat housing can be purchased, as well as many variations of the penning mentioned above. Tunnels constructed of a framework covered with polythene, a 'Polypen', can serve as winter shelter for fleece-bearing breeds. Field shelters of several types: corrugated iron arks, wooden stable-type shelters and fibreglass 'kennels' are all available commercially.

Plate 5 (p. 66) shows an excellent example of the loose-box system of housings, and a number of useful details. Note the insulation.

Figs. 6 and 7 illustrate a practice layout for goat yarding, designed by a goatkeeping architect with many years of practical experience. A yard of this type facilitates the feeding of bulky foods – a most laborious task

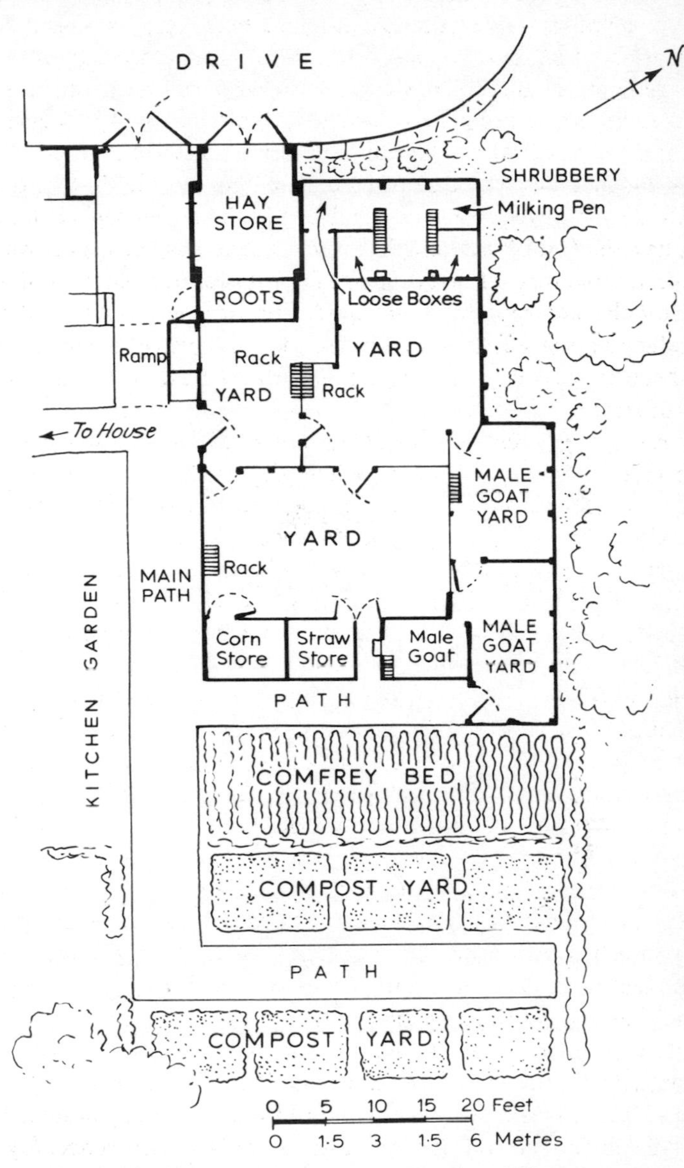

Fig. 6 Sketch plan of practical goat yards, Hanchurch Yews.

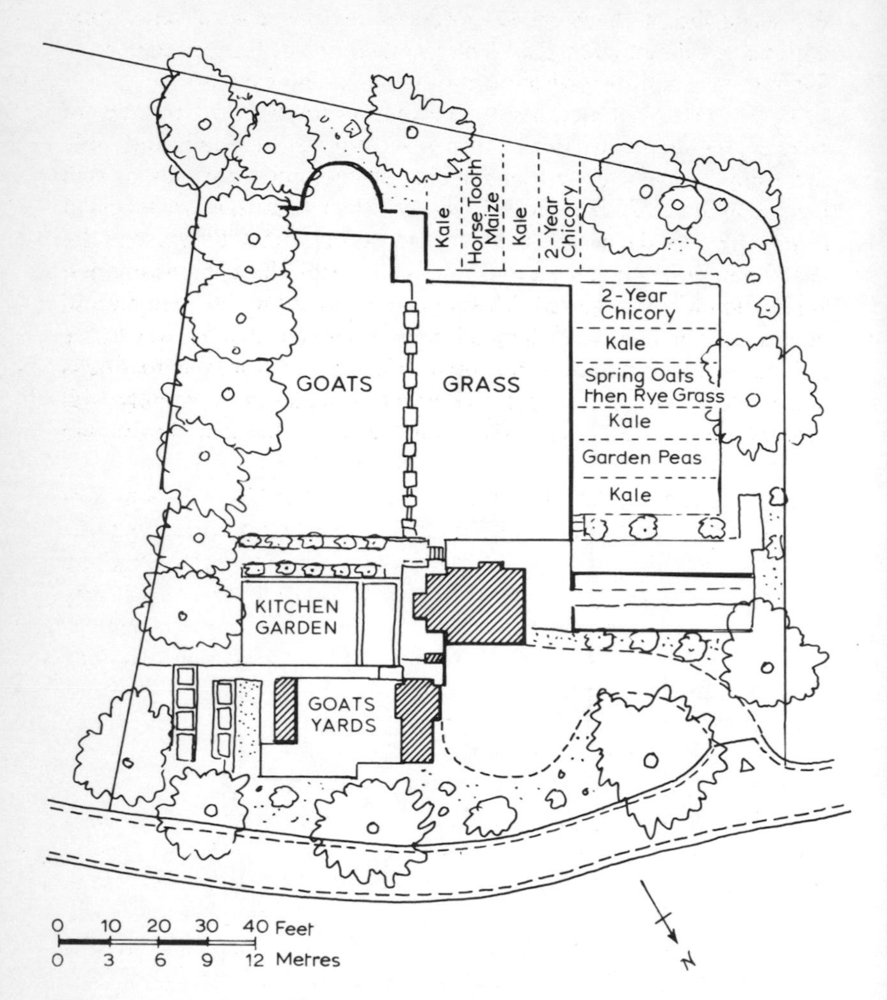

Fig. 7 Sketch plan of Hanchurch Yews showing position of yards and fodder crops.

when goats are fed in stalls or loose-boxes. It permits development of the social life of the herd, so necessary to good control, and quite impossible for hand-fed goats kept by other means. It keeps the hand-fed goat in a healthy environment of fresh air and exercise.

The system presented here is an unusual solution to a number of goatkeeping problems, which has stood the test of time and milk recording. It has every advantage over more popular systems of stall feeding and a great many advantages over the less satisfactory forms of free range – not the least of these advantages being that the whole of the goats' output of dung on a bulky diet is available for composting and application where required. Most of it is deposited in the open air and open spaces of the yards, where it is most easily collected. However, the milking pen would have to be completely shut off from the goat pens and made in such a way that it could be washed out, to comply with present recommendations, if the milk were to be sold for human consumption.

Fig. 8 illustrates an American-type of goat-feeding rack which is usefully versatile. With the locking bar lowered it becomes a feeding heck in which the goats can be isolated for concentrate feeding, and held for milking if required. With locking bar raised it becomes a free-for-all hay rack, in which the goat can nose around for the bits she most fancies but has great difficulty in wasting her second choice. The degree of isolation provided with the locking bar lowered is probably insufficient for feeding concentrates in large quantities and individual rations. But it is quite adequate for feeding the standard allowance of concentrates suggested in this book.

With a 3ft (90cm) feeding passage down the centre, this rack could be standard equipment in the communal goat house, the only other equipment needed being sleeping benches and drinking bowls. Nowadays feeding equipment of this type can be purchased ready-made, either as simple bars, or as a 'cascading yoke', usually incorporated into a milking parlour, where each goat's head becomes trapped in turn, then all can be released by pulling a lever. The food is placed in the passage outside the barrier. The most sophisticated farms have a moving belt to carry the forage along in front of the goats. Goats with side-spreading horns, such as Cashmeres and Angoras, cannot of course use barriers of this type, and hay has to be fed to them in ordinary racks, and concentrates in troughs. Hay nets are to be avoided, as they are dangerous.

If a higher degree of control of concentrate feeding is required, this

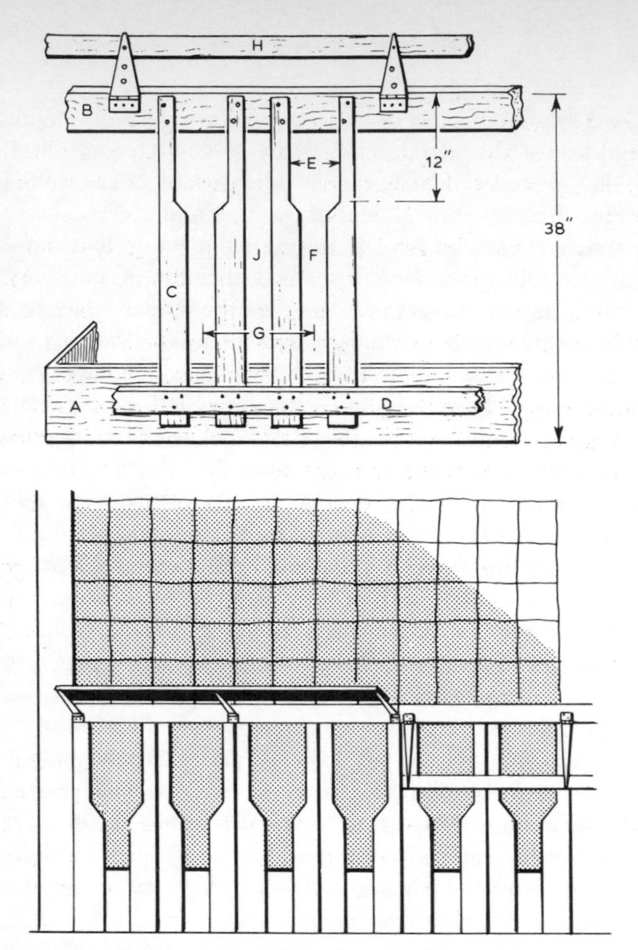

Fig. 8 The American-type goat-feeding rack (After I. A. Richards).

A Base board of the side of feed box.
B Top batten of feed box, 38in (96cm) from base A.
C 3in × 1in (8 × 2·5cm) stanchion – cut-out piece, 12in × 1½in (30 × 4cm).
D Foot rest, needed for goatlings and kids only to enable them to reach the head opening.
E Head opening: for kids, 6in (15cm); for milkers, 7in (18cm); for males, 9in (23cm).
F Neck slot: for kids, 3in; for milkers, 4in (10cm); for males, 6in.
G Distance between centres: for kids, 12in; for milkers, 18in (45cm); for males 24in (60cm).
H Batten on hinges, lowered to lock heads in the stanchions.
J Where J is more than 3 in, it is filled.

would be as well provided by a separate milking parlour.

Of the smaller items of goat equipment, the most important is the hay rack. The nature of the ideal goat hay rack is a subject of much controversy.

The wooden rack in general use is about 2ft (0·6m) deep, with wooden bars 1¼in (32mm) square, set 1½in (38mm) apart. The hay slides through them more satisfactorily if the bars are bevelled from 1¼in wide at the outside to 1in (25mm) wide on the inside. In making such a rack, screw the bottom of the bars *inside* the bottom rail of the rack, and the top of the bars *outside* the top rail of the rack. In loose-boxes it is an economy to place a two-sided rack on the pen partitions, to serve both occupants of the adjacent pens.

Iron hay racks of similar design are obtainable from the goat-equipment specialists.

It is necessary, and extremely difficult, to feed hay to Angora goats so that bits do not contaminate the fleece. A free-standing round rack has been suggested. With all goats, the saying that 'goats are accidents waiting to happen' comes tragically true in connection with hay racks more often than in any other way, mainly through heads and legs becoming trapped.

The rack is also very handy for feeding long, thick-stemmed foods such as kale, hogweed, artichoke tops and tree branches. These foods otherwise have to be chopped and fed in pails or *individually suspended*!

Drinking arrangements in goat houses appear to be rather primitive at first sight. Buckets predominate. They have to be emptied and refilled whenever a few hay seeds have contaminated the surface, for the goat is a fussy drinker of cold water; but she will nevertheless drink up to 5 or 6 gallons (22·7 or 27·3 litres) of water in a day – a lot to carry to one goat. Goats drink quite willingly out of automatic cow-drinking bowls, and the more old-fashioned, ever-flowing gutter system suits them well. But goats prefer and sometimes need to have their water with the chill off; moreover, such foods as sugar-beet pulp (soaked) come in buckets, and the miscellany of gruels which figure in the diet of high yielders also come in buckets – so buckets remain. Nevertheless, for a commercial herd, automatic drinkers and an electrically heated and themostat-controlled supply would be cheaper.

Goats are infernally inspired to be able to undo almost any kind of latch, including the simple bolt, which is an eternal nuisance to the attendant. As a contribution to goat-house equipment, fig. 9 illustrates, with more hope than confidence, a latch that is believed to be goat-

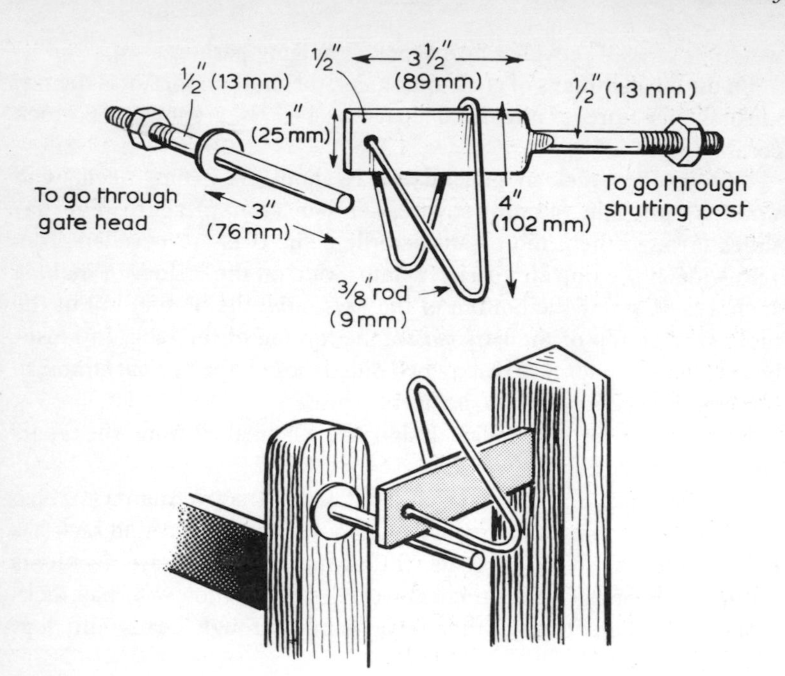

Fig. 9 A goat-proof latch.

proof. This is essentially a large-sized variation on a common latch which goats do defeat. Perhaps the 3in (76mm) projecting bar might be a little embarrassing in the more compressed loose-box houses, but it is excellent for main doors and gates, which have to withstand goats. A simple alternative is to add a second bolt near the bottom of the door in question.

CHAPTER SIX

The Principles of Goat Feeding

Basic principles of the goat's digestion

In our earliest days, nursery picture-books impressed it upon us that cows, sheep and goats grow fat living on grass. Yet later on, calorie-conscious, we learned to stay slim by eating green salads. It all seems so fundamental that many people may never stop to ponder on the small miracle of ruminant digestion. The goat by itself can no more digest, obtain energy and grow fat on the cellulose of plant tissues than we can; mammals do not have the necessary digestive enzymes. The secret of the goat's success, and the reason for man's association with all ruminant farm animals, lies in the rumen, the first of four 'stomachs'. Figure 10 shows the relative proportions of these in the adult goat. Only the last of them, the abomasum, has digestive functions similar to human stomachs. While all mammals, including man, have some cellulose-splitting bacteria in their intestines, the rumen has evolved into a vast fermentation vat, holding about two gallons in the goat, which teems with microbes: bacteria, fungi and protozoa of many species, which work together to ferment, in the absence of oxygen, the cellulose, starch and sugars in the food eaten.

These carbohydrates are broken down to yield Volatile Fatty Acids (VFAs), methane gas and some heat. The VFAs are absorbed through the rumen wall and combine with oxygen to act as the goat's energy source. The methane has to be voided by belching. Up to 25 per cent of the food's energy content is lost in the heat and the methane. The bacteria themselves obtain their energy for growth and multiplication from the fermentation process. The end-products of fermentation of the cellulose, starches and sugars differ in ways that have important practical consequences for goatkeepers.

Cellulose

This, the major component of forages, is the most important ruminant nutrient. It ferments completely, but more slowly than starch or sugars.

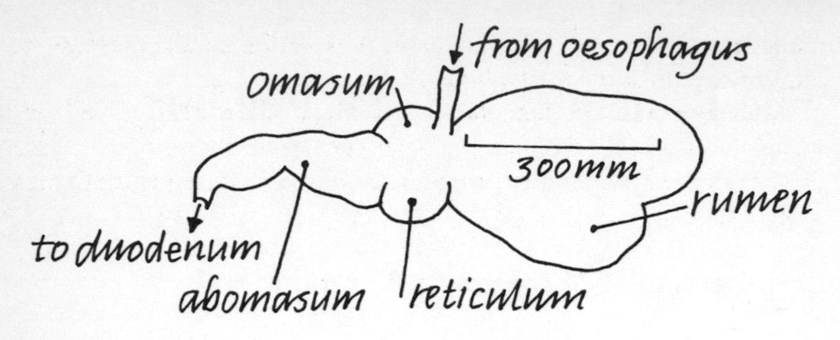

Fig. 10 The relative sizes of the 'stomachs' in the adult goat.

Unfortunately, from our point of view, plant stalks are strengthened by the woody substance lignin, which is not fermentable and protects the cellulose from bacterial action. Stalky forages are therefore, in effect, less digestible to the goat. The fermentation of cellulose produces mainly acetic acid, which supplies energy and milk butterfat. The bacteria which ferment cellulose are very sensitive to a build-up of acidity around them (i.e. if acids are produced more rapidly than they can be absorbed). The measure of acidity, pH, is 7·0 at neutrality; progressively lower numbers indicate increasing acidity while higher numbers show alkalinity. The cellulose-fermenting bacteria fail if the rumen pH falls to 6·0. Their growth-rate also slows or stops if too much fat is included in the goat's diet.

Starch

Starch and sugars could be digested without fermentation, but, since they enter the rumen, they are subject to microbial action. Starch, the main ingredient of cereal grains, is rapidly fermented by different varieties of microbe from those which attack cellulose. They are relatively insensitive to acidity but below pH 5·5 only a few types survive; one produces lactic acid, others turn the lactic acid into propionic acid. This, though important to milk yield, does not produce butterfat. If the population of the propionic-acid producers is small, for instance in a goat not used to concentrate feeds, then lactic acid may accumulate and be absorbed from the rumen in sufficient quantity to cause acidosis in the goat, which will consequently go off its food or even die. This is the main danger of over-feeding cereals. It is important to remember also that the acid produced after eating a feed of concentrates may well inhibit the activity of the cellulose-fermenters and thus reduce the goat's forage intake.

A diet of 100 per cent hay gives rise to a ratio of 3:1 acetic to propionic acid, whereas one of 60 per cent concentrates to 40 per cent hay gives a 1·5:1 acetic to propionic ratio, with consequent effects on milk butterfat level.

Sugars

These are present in molasses, roots, grass, good hay, etc., and, like starch, are quickly fermented to yield propionic acid, though some butyric acid may also be produced, which increases butterfat. As the feeds mentioned are generally eaten more slowly than cereal grains, acidosis is less likely to result.

Different foods are fermented at different speeds and remain in the rumen until, by chewing, fermentation or because of prior processing they become broken down enough to pass out through the reticulo-omasal orifice, which is small. Regurgitating the food from the rumen as cud and rechewing it helps in this process. Rumen acidity is partly neutralized by the alkaline saliva. Unfortunately the foods producing acid the most quickly, i.e. the concentrates, are the ones which need the least chewing so get less saliva mixed with them than forages. For the health of the rumen and the goat, at least 50 per cent of the diet, by weight on a dry-matter basis, should be in the form of forage.

While roots may be fermented in two hours, straw may take two days. Parts of some foods are indigestible and cannot be fermented at all. Straw, for instance, leaves much indigestible residue which has to be broken down by chewing the cud, slowing down its rate of passage and thus reducing the goat's food intake and hence lowering productivity. However, substances which remain for a longer time in the rumen are likely to be digested to a greater extent by the more prolonged microbial action to which they are thus subjected.

Proteins in the diet

The protein nutrition of ruminants, and their microbes, is very complex and still being researched. Much of the protein the goat eats is fermented by the rumen bacteria (and is thus known as Rumen Degradable Protein, RDP) to form acids and ammonia, which is a simple compound containing nitrogen, the essential element of all proteins. The bacteria obtain a little energy from this process, but much less than that derived from the fermentation of carbohydrates. Some of the ammonia is used by the bacteria to build their own proteins, but, particularly if they do not have enough energy to allow their rapid growth, surplus ammonia is absorbed by the goat and is converted into urea and excreted in the urine. Too much nitrogen in the goat's diet can lead to ammonia toxicity.

Alternatively, if there is not enough protein in the goat's food, bacterial growth and hence fermentation activity is slowed down, and the animal's food intake is therefore reduced. The balance of energy-giving foods and protein in the goat's diet is thus crucial to optimum fermentation rate and hence to the goat's health and productivity.

In all ruminants, some of the urea produced is recycled, via the blood and saliva, back into the rumen, where it is used to build microbial protein. There is evidence that this process occurs to a marked extent in goats.

Much of the protein which becomes available to the goat for digestion in the abomasum, and subsequently for body-building, etc., is the substance of the microbes, passed along from the rumen. It has recently been found in sheep that, instead of a given amount of a food producing a given amount of microbial protein, as previously thought, the amount of microbial protein available depends on the rate of passage of the food through the rumen, which in turn depends on the amount of food eaten relative to the size of the animal. A smaller sheep derives more microbial protein from a given amount of food than a larger one. This adds even more complexity to the estimation of an adequate diet.

When the needs of the goat for protein are very great, e.g. in early lactation, microbial protein is insufficient, and her diet must include some Undegradable Protein (UDP) which will pass, unfermented, through the rumen and on into the abomasum for digestion.

Fats and oils

Plant-derived feeds do contain fats and oils (known as lipids), particularly in the seeds from which oils are extracted. Since lipids dissolve in ether, the proportion of them present in a feed material is expressed in the composition tables as 'ether extract'. Lipid digestion starts in the abomasum, and a proportion of what is available will be microbial fat. Lipids are highly digestible and a good energy source. They also promote milk fat and a glossy coat; the total diet of the goat should contain between 2 and 6 per cent of ether extract.

Goats compared to other ruminants

Agriculturalists compare the goat's digestion and feeding habits with those of other ruminant animals not only out of biological interest, but because of the importance of such knowledge for effective animal performance, land management and forage utilization, and for conservation of the environment. Extensive, rather than intensive, livestock systems are seen to be increasingly important in the future agriculture of developed countries. In Scotland, the Macaulay Land Use Research Institute (MLURI), where ruminant grazing studies are much in evidence, produced the following comparative table in its 1988/9 Report.

Table 2 Characteristics of animals with different foraging strategies

Animal	*Cattle*	*Sheep*	*Goats*	*Deer*
Foraging strategy	Grazer	Mixed, preference for grass	Mixed, preference for browse	Mixed
Degree of selectivity while feeding	Low	High	High	Intermediate
Muzzle width (rel. to body size)	Wide, flat	Wide, flat	Narrow, pointed	Flat Intermediate
Rumen size (rel. to body size)	Large	Large	Small	Intermediate
Digestive capability:				
Fibrous grasses	High	High	Low	Intermediate
Trees, shrubs	Low	Intermediate	High	High

The goat as a mixed browser/grazer

It is an inescapable fact that goats love to eat browse (leaves and twigs), and this is as true of wide-ranging goats in many parts of the world as it is of back-garden-dwellers in Great Britain. So it can be assumed that the goat is naturally at least a part-time browser and we can expect to find evolutionary adaptations fitting in to this life-style, not found in entirely grazing ruminants. However these differences are merely a matter of degree; the goat is seen as part-way along a trend from browsers to grazers and, given the choice, it will eat grasses at times of year when these are at their best. To accommodate this mixture of eating habits, goats have adapted in a number of ways:

Eyes The rectangular pupil of the caprine eye is said to enable it to function well during rapid changes from light to dark surroundings and it has been suggested that this facilitates awareness of danger when the goat moves from sunlit patches into deep shade while browsing.

Behaviour The ease with which goats stand on their hind legs, stretching up with mouths and even fore-limbs to bring branches within reach, increases greatly the availability of browse; this habit is not seen so well in grazing ruminants. Furthermore, it has been noted that goats will travel considerably farther in a day than sheep to seek out food to their liking.

Muzzle The narrow, pointed and mobile muzzle is perceived as helping the goat to pick off with accuracy and consume only those fragments of a plant which have been selected as desirable, rejecting the remainder. The goat is quite a small-sized ruminant. As animal size decreases the relative amount of energy required for body maintenance increases. It is therefore vital for small ruminants such as the goat to be able to select and consume the most nourishing parts of the fodder available.

Sense of smell How does the goat exercise this high degree of selectivity? Very few experiments on this subject have been done; results indicated differing responses in sheep and goats, and yet sheep also have considerable selective powers. This is one of many areas in which further research is needed.

Mouth No one who has watched a goat scraping bark with the front teeth of the lower jaw, or carefully mouthing thistles and nettles so as to eat them while remaining unscathed, then later grinding the regurgitated cud between the molars, can doubt the perfection of the goat's mouth for its purpose.

Rumen Browsing animals are typified by having a smaller rumen, in relation to body size, than grazers. The fact that they can consume a great deal of food is due to the more rapid rate of passage of that food into the abomasum than occurs with a grass diet. Browse is a mixture of rapidly fermentable leaves and indigestible lignified fractions; microbial action soon extracts the readily available nutrients and there is then little point in the indigestible fragments remaining in the rumen when their voiding will leave space for further ingestion of leaves. Browsers tend to have a larger orifice from the reticulum to the omasum than grazers. A quick throughput from a small rumen also maximizes the arrival in the stomach of plant cell contents and of microbial protein. The caprine rumen is long and narrow compared with that of the sheep. The internal surface area, important for rapid absorption of the acids produced by fermentation, is expanded by leaflike papillae.

This picture is confused by the appearance of statements by some authors that the rumen of the goat is larger than that of the sheep. This is thought to be due to individual differences caused by diet; the goat is, after all, the most adaptable of ruminants. Grazing ruminants as a whole are typified by a large rumen containing a slowly, but more completely, digesting mat of grass and bacteria. Browsing animals do not achieve success as grazers of grass alone. Browsing would appear to provide the high energy diet needed by small ruminants. If browsers are large they need access to a lot of good quality food; the giraffe is the extreme example!

The goat/plant relationship

It must be recognized that plants have evolved too, and often have survival characteristics which prevent them from being palatable food for all and sundry; they may in fact be downright poisonous. Selectivity is important for browsers; goats are renowned for selecting the most nourishing portions of plants; if they do find themselves confined to stalky grass they will eat seed-heads if they can. However, another important caprine success-factor is the willingness to try a wide range of browse-plants and the ability to tolerate bitterness of flavour and plant chemicals which many other animal species cannot. (Alas, in the unnatural surroundings of a domestic garden, this inquisitiveness frequently leads to goats' being poisoned.) As an example of the superior digestive ability of the goat, it has been shown that willow and gorse have a much greater nutritional value to goats than to sheep; it has been suggested that proteins in the goat's saliva can bind and nullify the plant tannins, which otherwise inhibit digestion.

But there are many species of unpalatable plant and within one species, variations in palatability occur. Plants growing in sunshine, but only slowly, perhaps due to dry conditions, are the most likely to have an excess of the products of photosynthesis which will then form unpalatable or poisonous compounds.

Goats on a diet of poor quality forage

The ability of goats to select the most nourishing fraction of plants enhances their ability to survive if they are faced with a poor diet such as cereal straws. Work at Reading University has shown that the parts of the straw eaten, the leaves and leaf sheaths, are higher in nitrogen and digestible organic matter than the parts rejected, the stems. The leaves from straw have, in fact, a digestibility akin to that of hay. As the leaf and sheath fraction of straw represents only about half its weight, it follows that twice as much must be offered as the goat can reasonably be expected to eat, and a rejection rate of 50 per cent must be accepted.

It has been observed that adult goats digest poor quality, i.e. high lignified fibre, low protein, forage better than adult sheep, although on medium and good quality diets both species do equally well. This observation may be partly explained by the selective ability of the goat, but sheep are also selective feeders. Actual differences in the goat's digestive process have been sought, and found, but results of different experiments are conflicting and further work is needed; research on this subject is ongoing in several countries.

One factor is the greater salivary secretion in goats than sheep (29·8 fl oz (848ml) compared to 17·6 fl oz (502ml) per day in one study). Urea from the blood is recycled via the saliva, as mentioned above, increasing the nitrogen supply to the rumen microbes, optimizing the fermentation rate and increasing the amount of microbial protein available to the goat.

The length of time which the food takes to pass through the gut is intimately bound up with the degree of digestion which takes place, a slower passage increasing the digestibility of poor forage. Other factors influencing retention time are the amount eaten and the drinking frequency, increases in both of which speed the passage of food. There is some evidence that the passage of straw through the goat's entire digestive system may take longer than that in sheep, allowing a greater extraction of nutrients. Experimentally, goats have been said to drink less than sheep and when adult to eat more of a straw diet. This latter seems difficult to reconcile with a higher retention time. The variation in reported results is probably due to different experimental circumstances, for instance housed or free-range animals.

Differences between the feeding habits of sheep and goats have been reported. Goats take more frequent meals than sheep, and when exercising their selective powers to a high degree will take longer to consume the same weight of feed than sheep. Goats chew the cud for relatively less time than sheep; on the other hand it has been reported that goats eating wheat straw have more small particles in their oesophageal contents than sheep.

Reported evidence of better rumenal digestion of poor quality forage in goats than sheep includes: a higher concentration of bacterial protein in the rumen of goats than sheep; a higher concentration of cellulose-splitting bacteria in goats. On a diet of only sodium hydroxide-treated straw the bacterial population in the goat's rumen remained constant while that in the sheep's rumen fell; some workers have found a higher VFA concentration in the rumen of goats than sheep on poor forage; supplementation of the diet with soya-meal removed the difference between the two species without increasing the bacterial activity in the goats. But, protein supplementation of a poor forage diet has been found to increase uptake more in goats than sheep, and in spite of the evidence mentioned some experts maintain that there are no major differences in digestion between goats and sheep. The results of future research on goat digestive capabilities will help to answer many questions. Meanwhile breed differences have been, not surprisingly, discovered: for instance, the Black Bedouin goat was found to digest wheat straw with 13 per cent greater efficiency than the Saanen.

Assessing the goat's needs

Up to now this chapter has painted a picture of the goat as a small ruminant, adapted by evolution to select, live and reproduce on a varied diet of highly nutritious browse, but able to include good quality grasses in its diet. The continuing existence of feral herds (at least until they become the target of the 'sportsman's' gun) proves that such an existence is possible not only in warmer climes but here in the UK.

The domesticated goats of today's developed countries, however, have been bred-up to a high degree of productivity, of milk, flesh or fleece, which requires supplementation of the diet with foods which yield concentrated nourishment. Such supplementation must be done with great care, never once losing sight of the basic principles of digestion; there are two sufferers if we make a mistake, the goat and our bank balance.

Those who keep highly productive animals, particularly housed goats, have shouldered the considerable burden of responsibility for providing every necessary item of the diet. In this Age of Legislation, our moral obligation to our charges has been reinforced by law: The Welfare of Livestock Regulations 1990 (SI 1990 No. 1445) requires goats which are in any way part of an agricultural business to be fed a wholesome diet in sufficient quantity to maintain them in good health and to satisfy their nutritional needs, and to have access to food each day. In addition, the MAFF Code of Recommendations for the Welfare of Goats (PB0081), which came into operation on 1 August 1989, and breaches of which can be used in evidence against those being prosecuted, gives a brief but accurate summary of the main food requirements of goats.

The law should create no conflicts of interest; it is good business to provide each member of the herd with a ration formulated to ensure maximum health and productivity, while doing so at the lowest cost, for the size of the feed bill is usually the most crucial factor in the success or failure of the enterprise.

Two sets of information are required for this formulation, in compatible terms: the varying needs of the goat, and the nutritive value of the available foodstuffs. The two major concerns of the goatkeeper are supplying the energy and protein needs of the herd.

Energy

The goat requires energy for maintenance of body functions, temperature control, activity, growth and all types of production. Some needs can be met by using the energy reserves stored in the body fat, provided that these are adequate beforehand and replaced later; dietary energy allowance must be made for this, typically in late lactation. Energy requirements are measured in the laboratory with goats housed in respiration chambers where heat production can be measured.

Growth

The energy requirements for body weight increase are confused by whether fat or lean (muscle) is being laid down. Lean contains 80 per cent of water, and fat very little water and more energy than lean; for a given weight increase up to eight times more energy is required if the tissue being deposited is fat than if it is muscle. The amount of feed required for a unit of weight gain is described as the 'feed conversion'; this will tend to be considerably higher in older animals laying down fat than in young ones increasing their muscle weight, and varies also with the type of feed given. Precise data on this subject for goats requires further experimentation.

Lactation

Energy requirements for milk production are relatively high, and complex to determine as they are superimposed on a variety of other requirements, such as those for body maintenance, maybe growth, weight gain or weight loss, maybe pregnancy, and milk composition, particularly the fat content. This latter point is the reason why the energy needed for lactation is expressed per unit of 'fat-corrected milk'. The standard taken is often 4 per cent fat. If the milk secreted has a fat content of more than 4 per cent, the volume of fat-corrected milk will be greater than actual; if less, then the corrected volume will be less than that actually produced.

Foods

Energy and other values of feedstuffs are always compared on a 'Dry Matter' (DM) basis as water contents vary greatly, between mangolds and hay for instance. Some parts of foods are indigestible; these too are eliminated before comparisons are made on the 'Digestible Dry Matter' (DDM) content. A further refinement is made by eliminating the mineral content of the feed, which though valuable, does not provide energy. If the food is burned at high temperatures, the mineral content is what remains; it is therefore sometimes referred to as the 'ash' content. The part that burns is the Organic Matter. 'Digestible Organic Matter' (DOM) is calculated by discovering the ash content of the food and the resulting dung. 'Digestible Organic Matter in the Dry Matter (DOMD) is a useful comparative figure found in feed tables.

The digestible energy of a foodstuff can be determined by burning it in the laboratory and measuring the heat produced, then doing the same with the dung produced after eating that foodstuff, and subtracting the latter value. But, when the animal concerned is a ruminant, energy is lost during microbial fermentation of the food, as heat and in methane production. So to measure the energy value of a food to a ruminant the term 'Metabolizable Energy' (ME) is used; this is determined by subtracting the energy in the methane, and also in the urine produced, from the digestible energy. Experiments to discover the energy in the methane gas produced by the goat require expensive equipment so the ME values of many feeds are calculated from experiments made using similar foodstuffs. The term 'Net Energy' (NE) is sometimes used.

The old Starch Equivalent system was a Net Energy system. It is a measure of the extent to which ME is used by the goat for different purposes such as milk production or maintenance. It is a useful concept but complicated by the fact that the NE of a food varies with the use to which the energy is put by the goat. So ME is the term in common use, the idea being that the goatkeeper can look up the needs of goats of different weights and at different stages of the reproductive cycle, and the ME values of different foods, and match the two together. No one will be surprised that successful goat feeding is not as simple as that!

Some of the problems are: the composition of lean or fat of weight gain is not precisely known and, as stated above, makes a big difference to the energy needed. The existing condition of the animal, for instance, will influence whether or not fat is laid down; the acids produced by fermentation of concentrates may inhibit the digestion of cellulose so that when feeds are mixed the anticipated energy value may not be realized. Also, the extra energy for activity required by goats is not precisely known nor is it possible to predict food intake very accurately as yet.

Nevertheless, the ME system has been in use in the UK since 1976 and is a useful guide to correct feeding. ME is usually given in Megajoules (MJ). When describing the needs of the goat, the MJs per kilogram of 'metabolic weight', rather than liveweight, is often used. Metabolic weight is liveweight to the power 0·75 ($W^{0.75}$, or $\sqrt[4]{W^3}$ where W is liveweight). This method irons out differences due to the size of the animal, smaller animals having a relatively higher metabolic rate. When describing the energy content of the feed, remember that the MJs per kilogram of the dry matter of the feed are given. These figures are for adult goats:

Maintenance energy

An average figure obtained from a large number of trials of housed goats is 0·445MJ ME/kg $W^{0.75}$/day.

Additional energy for milk production

4·95MJ ME/kg of milk of 4% fat, plus 0·078MJ ME for each 0·1% over 4%.

Energy requirements in pregnancy (includes maintenance)

0·6MJ ME/$W^{0.75}$/day, six weeks before kidding, rising to
0·8MJ ME/$W^{0.75}$/day, one week before kidding. (W is liveweight in kilograms.)

Energy gained from loss of liveweight

This is obtained from mobilization of the fat reserves and varies in different reports, but a typical figure is 28MJ/kg weight lost. The energy content of the diet in early lactation is important in limiting this loss.

Energy necessary for restoring body reserves

Most goats in milk lose weight in early lactation and need to replace their reserves after peak lactation has been reached and passed. Allowance must be made for this in the diet. Once again authors vary; two figures suggested are 34MJ ME/kg liveweight gain and 52MJ ME/kg.

Energy requirements of young goats

The maintenance energy, i.e. without growth occurring, can be influenced by the method of weaning, but may be a little more than for adults. For the Granadina breed in Spain, figures obtained were:

0·443MJ ME/kg $W^{0.75}$/day, during milk feeding,
0·557MJ ME/kg $W^{0.75}$/day, during the period just after weaning,
0·427MJ ME/kg $W^{0.75}$/day, during ruminant period.

The additional energy required for growth varies with the weight of the kid, the rate of growth, breed, feeding conditions etc.: successful size increase of kids is heavily influenced by their energy intake. Initially, abomasal development limits food ingestion, but between two and four weeks of age the energy content of the diet begins to be the most important factor. Also the intake level is influenced by the stage of growth. A combined figure given for growth and maintenance is:

0·837MJ ME/kg $W^{0 \cdot 75}$/day.

Alternatively, energy requirements have been expressed per 100g of gain in weight, ranging from 1·3 to 1·5MJ for milk-fed animals and from 2·15 to 5·9MJ for ruminating youngstock, the higher figures being found when less energy-rich diets were fed. Thus diets with a high 'energy density' are necessary for good growth in kids.

Energy requirements for activity

The figures given above are for housed goats. The additional requirements for activity are not precisely known, but a rule of thumb is to allow 25 per cent extra energy for animals grazing on lowland, and 50 per cent extra for animals on hill grazing – not inconsiderable amounts.

Energy requirements for fleece production

Once again, there is not yet very precise knowledge of this matter. However, for cashmere goats, which are producing under-down between late June and late December, assuming a down weight of 250g, and a total fleece weight, including the guard-hairs, of 612g, the recommended energy allowance is 0·084MJ ME/day, for the six months of the down-growing period.

Angora goats produce a considerably greater weight of fleece – several kilograms – which grows throughout the year. It is suggested that an allowance of 0·25MJ ME/day be added to the diet for every 2kg of anticipated fleece production per year, which is fairly close to the recommended allowance for Cashmere goats; this extra energy should be added to the ration throughout the year.

Protein requirements

The goat needs protein to supply the rumen microbes, to replace and build body tissues and to produce young, milk and hair. Proteins are made up of 'building blocks', the essential amino-acids. These have to be supplied in the food except that, as described earlier, in ruminant animals the rumen microbes manufacture proteins during the course of their growth, which is later available to the goat. The bacteria need a supply of proteins, or simpler nitrogenous compounds such as urea, from which to construct their particular proteins; the goat may need more proteins than the microbes can supply, in the diet. The fat reserves of the body contain no protein; if these are being utilized a protein source has to be found.

All proteins are not digestible, so the term 'Digestible Crude Protein' (DCP) is often found in tables of feed values. Due to the complexities of microbial fermentation, however, systems have been proposed which can more accurately predict the food-protein/ruminant relationship. In the last ten years much research has attempted to determine the goat's nitrogen requirements. Subjects studied have included the breakdown of protein in the rumen and the synthesis of new proteins by the rumen microbes; their requirements for degradable protein (RDP), the recycling of nitrogen as urea, the efficiency of digested proteins for milk protein synthesis and the mobilization of the body's protein reserves. All of this work has been directed towards the more accurate estimation of the goat's needs for dietary proteins, always the most expensive part of the diet. There is still much more research to be done.

As mentioned earlier, the balance of RDP, UDP and energy in the diet is very important. Too little RDP in relation to the digestible organic matter will reduce rumen fermentation and hence the goat's productivity.

Protein for maintenance

This has to cover losses in faeces, urine, hair and scurf (shed layers of skin). From experimental work, a requirement of:

2·5g DCP/kg $W^{0.75}$ has been suggested, but there are widely varying estimates, depending on the quality of protein in the diet.

Protein for lactation

The efficiency of converting digested protein to milk is higher with good quality protein, that is protein rich in the essential amino-acids, such as fish-meal, but the amino-acid requirements of dairy goats are not known in sufficient detail. As a rough guide, goats in early lactation have been found to need a diet containing 18 per cent Crude Protein (CP), in mid-lactation 16 per cent CP and in late lactation, 12 per cent CP. In early lactation some of the protein reserves of the body may be mobilized and used to produce milk protein; an estimate is that 230g of milk protein could be produced in this way during the first three weeks of lactation. Approximately 50g of DCP are needed for each kg of 3 per cent protein milk.

Protein for growth

Protein is needed for laying down muscular tissue, but not for fat, and so knowing the composition of the increase in body weight is important to assessing the protein needed to make that increase. Our knowledge of that composition comes from the study of carcasses. In general, as an animal gets older its increase in weight is more fat and less lean. It has been shown that kids take up relatively more nitrogen at two to six weeks of age than at nine to thirteen weeks. The estimates of protein requirements for growth made by different research workers vary considerably. These differences could be due to a number of factors such as breed, growth rate, body condition and composition, sex, nitrogen and energy levels of the diets and experimental methods. An average figure obtained is that of:

0·28g CP/g of bodyweight increase.

This is in addition to maintenance requirements. This corresponds to 0·195g of DCP. A good working rule is to provide growing kids with a ration containing 18 per cent of the dry matter as crude protein. Less than this has been seen to result in higher feed-to-gain ratios, and there is evidence that the high-protein diet can be continued with advantage up to five months.

Yet if all of this protein is degraded by the rumen microbes (RDP) the potential benefit will be lost as nitrogen compounds excreted in the urine, and the extra expense will be wasted. Protein evaluation systems indicating RDP and UDP in the ruminant diet are being actively sought. Protein breakdown in the rumen does not only depend on the substance itself (e.g. fish-meal vs. soya-meal) but on the rate of passage through the rumen which in turn depends on many other factors, both animal and environmental. Continuing research is urgently needed.

Daily food intake

The energy and protein requirements of each goat have to be fitted in to the amount of food that the animal can consume in a day. This amount varies in mature females with the different stages of the annual cycle of pregnancy and lactation. It is the Voluntary Dry Matter Intake (VDMI) which has to be considered, and most of the experimental figures obtained are from housed dairy goats.

From about the second week of pregnancy, VDMI falls by about 300g/100kg of liveweight/month. The mean DMI of 393 French Alpine goats was 1·52kg/day during the last six weeks of pregnancy. In the last two weeks the fall in DMI is successively greater in goats carrying increasingly higher numbers of kids, but the biggest change in VDMI is during lactation, when it starts to rise soon after parturition to reach a maximum between the sixth and tenth weeks of lactation. In the first few days after kidding, before the appetite increases, milk production is largely dependent on the use of body reserves.

In French Alpine goats, a VDMI as high as 3·6kg/day, equal to 6·8 per cent of the goat's weight, or 181g/kg $W^{0·75}$ has been recorded. However a VDMI of 2·1kg/day, about 3·5 per cent of body weight, was the average figure. The rise is most rapid in the first four weeks after kidding, paralleled by the milk yield. After reaching its highest level, VDMI falls gradually by about 100g/month as milk yield falls.

In calculating the goat's dietary needs it is necessary to know her weight and how that changes during the year, also her milk yield, which has the greatest influence on her intake, in order to be able to predict how much food, measured as dry matter, she is going to eat at various times. In France, where somewhat complex equations have been developed for predicting the VDMI of goats, it was found that the individual differences between the food intake of lactating goats was due 50 per cent to milk yield and 30 per cent to body weight. Another factor influencing VDMI is the amount of fat stored in the body; substantial amounts of fatty tissue in the body cavity cause a marked reduction, during both pregnancy and lactation, in the capacity to take in food.

The quality of forage also influences intake, and too little is known about the intake and digestibility relationship of browse consumed by free-ranging goats. It is probable that, as with forage fed to housed goats, more is eaten when sufficient is available to allow a high percentage, up to 50 per cent, of the forage on offer to be selectively refused. On the other hand, less forage is eaten when concentrates are fed in substantial quantity; this is known as the substitution effect, and is particularly marked with starchy concentrates fed at high levels. Once the diet contains sufficient energy for the goat's needs, a further increase in energy content, sometimes described as the energy density of the diet, can reduce the goat's dry matter intake.

The diet previously consumed also affects VDMI, particularly around the time of kidding and for the first two months of lactation. A goat which has been encouraged to eat plenty of good hay in late pregnancy will eat more food after kidding, presumably because a bulky diet keeps rumen volume up while it is under pressure from the enlarged uterus. On average, for each 100g of hay eaten per day in the pre-kidding diet, an extra 45 grams of VDMI will be seen in early lactation.

The dry matter intake of a goat is therefore affected by many factors. In France an attempt has been made to quantify the effects of the factors by applying the 'Fill Unit' system, developed for dairy cows, to goats; it is felt that dairy goats respond to nutrition in a similar way to dairy cows. This system does not so far seem to have attracted British goatkeepers, who prefer to base their dietary calculations on a dry matter intake of 3·5 per cent of body weight with a rise of 5 per cent in early lactation.

This approximation is likely to be refined in future. Government funded research, the Technical Committee on Response to Nutrients, (TCORN), has reported on goat nutrition, and points to the prediction of food intake as one of the areas where further knowledge is needed.

Minerals in the diet

The minerals listed in Table 3 are said to be essential to goats, in that a deficiency of any of them causes a metabolic disorder which can only be prevented or cured by supplying the missing mineral.

Minerals occur naturally in all the feedstuffs that the goat eats, but it is necessary to provide extra in some cases, notably common salt

Table 3 Minerals (with their chemical symbols) essential to goats

Major minerals		*Trace Minerals*			
		I		II	
Calcium	Ca	Iron	Fe	Chromium	Cr
Phosphorus	P	Iodine	I	Tin	Sn
Magnesium	Mg	Copper	Cu	Vanadium	V
Potassium	K	Manganese	Mn	Fluorine	F
Sodium	Na	Zinc	Zn	Silicon	Si
Chlorine	Cl	Cobalt	Co	Nickel	Ni
Sulphur	S	Molybdenum	Mo	Arsenic	As
		Selenium	Se	Lead	Pb

The trace minerals in column I are those found most likely to require adjustment in the diet of goats in the UK.

(sodium chloride, NaCl). It must be remembered however that too much of any one mineral in the body can be just as damaging as too little, so, as with energy and protein, it is necessary to try to match the goat's particular and changing needs with the amount supplied in the diet. This is complicated by the fact that minerals interact in the body, so that an excess of one may create the need for an excess of another to balance it. An example is the way that calcium and phosphorus intakes must be carefully balanced.

Major minerals

Calcium Almost all the calcium in the body – 99 per cent – is stored in the bones and teeth, where it is the main component of their structure. The other 1 per cent is in the blood and tissue fluids, where its presence is crucially important in the conduction of nerve impulses, the contraction of muscles and the clotting of blood to seal wounds. The calcium in the bones forms a reservoir from which the level in the blood can be topped up; this delicate process is controlled by calcitonin, the hormone secreted by the parathyroid glands, and Vitamin D.

At kidding, there is a sudden huge demand for extra calcium for milk; if calcium does not arrive from the bones quickly enough hypocalcaemia, 'milk fever', results. Funnily enough, the very worst thing is to give the goat lots of calcium before kidding, as this will result in more storage in the bones, rather than mobilization from the bones. Calcium and

Plate 6 A British Boer Goat, young female. (Photograph: Michael Gaisford)

phosphorus should be present in the diet in the ratio in which they occur in the bones, Ca:P, 2:1. As a general rule, green foods and sugar beet pulp are high in Ca and low in P; concentrates (cereal grains) are high in P and low in Ca.

Many of the figures for mineral requirement have been taken from experiments with sheep because the work has not yet been repeated on goats. It must be remembered that only a certain amount of the mineral present in the feedstuff will be absorbed by the goat. For calcium, the amount absorbed is said to be 30 per cent of that in the food. Making allowance for this, the dietary content per day should be:

Calcium for maintenance: 0·067g/kg of liveweight: 3·4g for a 50kg goat.

Additional calcium during last two months of pregnancy: 4·2g/day (for 1·5kg gain/week).

Additional calcium for lactation: 4·2g/kg of milk.

Additional calcium for growth: 3·5g/100g of increase in body weight.

Phosphorus The major deposits of this mineral – 75 to 85 per cent – are in the bones. Of the remainder, phosphorus is a component of many enzymes concerned in the metabolism and the storage of energy. It is also a part of the genetic material and is regulated similarly to calcium. Deficiency can lead to fertility problems and unusual eating behaviour (pica). The feeding of males, particularly castrated males, must be carefully considered with respect to phosphorus, as an excess can cause blockage of the urinary system by stones. Basically this occurs in animals fed on concentrates but not in those fed on grass and other forages. If concentrates must be fed, calcium chloride can be added to correct the Ca:P ratio.

It is said that 65 per cent of phosphorus in food is absorbed. The daily diet should contain the following amounts of phosphorus:

Phosphorus for maintenance: 0·046g/kg of liveweight: 2·3g for a 50kg goat.

Additional phosphorus during last two months of pregnancy: 1·1g/day (for 1·5kg gain/week).

Additional phosphorus for lactation: 1·5g/kg of milk.

Additional phosphorus for growth: 0·92g/100g of increase in body weight.

Magnesium Of the magnesium in the body 60 to 70 per cent is in the bones, the remainder is in the soft tissues and body fluids, a component of many enzymes and important in the functioning of nerves and muscles. Unlike the situation with Ca and P, the magnesium in the bones does not form a reservoir controlled by hormones. The requirements have to be met by the diet and, like Ca and P, Mg is secreted in the milk. Staggers or hypomagnesaemia can result in high-yielding milkers, on lush grass, low in Mg, particularly if they are stressed in any way. Magnesium is therefore included in mineral feed supplements.

Magnesium for maintenance: (allowing for only 20 per cent absorption from the food) 0·017g/kg of liveweight: 0·85g for a 50kg goat.

Additional magnesium during last two months of pregnancy: 0·2g/day (for 1·5kg gain/week).

Additional magnesium for lactation: 0·7g/kg of milk.

Additional magnesium for growth: 0·2g/100g increase in liveweight.

Potassium and *Sodium* Potassium is found within the cells of the body, while sodium mainly lies in the fluids outside the cells. Their roles are concerned with the fluid balance of the body and the transmission of nerve impulses. They are not stored in the body, but the regulation of the amount excreted in the urine is the function of the hormone aldosterone, secreted by the adrenal cortex. Both cereals and grass tend to be low in sodium, and deficiency in salt leads to cravings, pica, licking walls, drinking urine, etc., also impaired appetite and digestion, reduced growth and milk yields, and poor reproduction.

Sodium is secreted in the milk so lactating females have particularly high needs. Salt is generally provided in the form of free-choice lick blocks, which are usually consumed enthusiastically by the herd. Excessive salt added to the diet is said to limit growth, and an excess of potassium, which tends to occur in goat rations, can hinder the body's ultilization of magnesium. Absorption of K is 90 per cent, Na is 80 per cent, from the food. For maintenance:

0.05g/kg of liveweight of potassium: 2·7g for a 50kg goat.
0·018g/kg of liveweight of sodium: 0·9g for a 50kg goat.
Additional during last two months of pregnancy: K 0·3g/day; Na 0·2g/day (1·5kg gain/week).
Additional for lactation: K 2·3g and Na 0·5g, both per kg of milk.
Additional for growth: up to 31kg body weight: K 0·26g/100g gain,
Na 0·2g/100g gain,
over 32kg body weight: K 0·044g/100g gain,
Na 0·05g/100g gain.

Sulphur Sufficient is provided in the protein, though supplements may increase fleece weight.

Table 4 Daily dietary allowances of major minerals

	Ca	*P*	*Mg*	*K*	*Na*
Maintenance (g/kg liveweight)	0·067	0·046	0·017	0·05	0·018
Additional in last 2 months, pregnancy (g)	4·2	1·1	0·2	0·3	0·2
Additional for lactation (g/kg milk)	4·2	1·5	0·7	2·3	0·5
Additional for growth (g/100g gain)	3·5	0·92	0·2	0·26	0·2
(over 32kg liveweight)				0·044	0·05

Trace minerals

These are so called because, compared with the major minerals, only tiny traces of them are required in the daily diet; nevertheless they are essential. Table 5 gives the recommended daily dietary allowances of trace elements.

Table 5 Daily dietary allowances of trace minerals, in mg/kg of food dry matter

Iron	30–40	Manganese	40
Copper	8–10	Zinc	50
Cobalt	0·1	Selenium	0·1
Iodine	0·4–0·6	Molybdenum	0·1

Iron This is important in the transport of oxygen round the body in the blood, but is stored in the liver, spleen and bone marrow. The amount in the body is controlled by a limit on the amount absorbed. Iron deficiency, or anaemia, may be encountered in three-to-eight-week-old kids whose diet is low in iron, or in older goats heavily infested with blood-sucking parasitic worms.

Copper This is also an important component of enzymes concerned with tissue respiration. It is also present in the black pigment melanin, and is involved in the formation of red blood cells. The amount absorbed from the food is influenced by many factors, one of which may be the animal's needs. Excess is stored in the liver. Copper deficiency in kids causes swayback; there seem to be genetic differences between breeds in the susceptibility to copper deficiency – possibly Pygmy goats are most susceptible. Goats do not appear to be as likely as sheep to suffer from copper poisoning, though there is a possibility that Angora goats are more susceptible than the dairy breeds; copper poisoning was seen in Angora kids fed on milk replacer containing 10mg copper/kg dry matter. Excessive molybdenum creates copper deficiency.

Cobalt The importance of cobalt is that it is required by the rumen microbes in the manufacture of vitamin B_{12}, which is needed by the goat for its nitrogen, carbohydrate and fat metabolism. The widespread feeling among goatkeepers that goats need more cobalt than sheep has

Plate 7 British Toggenburg kids in the Kinloch Herd nibbling a molehill in search of minerals.

not so far been borne out by science, though deficiencies have been recorded.

Iodine This is a constituent of the thyroid hormones, and 70 to 80 per cent of the iodine in the body is in the thyroid gland. In the unborn and young goat, iodine influences development; in the adult, basal metabolism and fertility. Surplus iodine is excreted in the urine. In the kid, iodine deficiency results in increased size of the thyroid gland; also slow growth and reduced vitality. In the adult, it reduces appetite and conception rate and causes abortions. Kids may be born hairless. Certain feeds, clovers and brassicas, tend to cause goitre, and to counteract this effect, extra iodine should be given: up to 2·0mg/kg of dry matter fed.

Manganese This has a wide range of roles in the body processes, including oestrous activity. Only 1 to 4 per cent of the manganese in the diet is absorbed, and ruminants do not appear to have reserves of this element. Severe deficiencies have been seen to affect reproduction; less severe ones, the Mn content of the milk.

Zinc This also has a wide range of functions; it is a constituent of enzymes, and of insulin; it has a close functional relationship with vitamin A and with the growth hormone, prolactin. Also it is involved in the healthy growth of skin and hair. To a certain extent absorption is governed by need. Surplus zinc is got rid of in the droppings. Zinc deficiencies in goats have been induced experimentally, but are also believed to occur naturally more often than is realized. As with other trace elements, an excess of zinc causes poisoning.

Selenium This is required only in minute quantities, but is very important in fat metabolism. Its action is closely related to that of vitamin E. Selenium is also thought to increase fertility and resistance to infection. A deficiency is said to occur often in kids, causing them to do badly and also to suffer from white muscle disease. Organic compounds of selenium are utilized better than inorganic ones, and are thus more likely to give rise to a toxic excess.

Molybdenum This occurs in the skeleton, liver and kidneys and, like other trace elements, is a component of enzymes. It is involved in the formation of uric acid. There is an antagonistic relationship between the compounds of sulphur and those of molybdenum. Naturally occurring molybdenum deficiency is almost unknown, but an excess can be serious as it causes a secondary deficiency in copper.

It would appear from all this information that formulating a goat diet which contains the correct minerals is a very difficult matter. Fortunately, this is not the case. Goats which are properly fed on a varied diet, balanced in energy and protein, particularly those which have access outside to browsing and grazing of deep-rooted 'weeds' and are fed hay containing plants other than grass alone, provided they have constant access to a salt-based mineral lick and a have a suitable commercially available mineral supplement in their concentrates, rarely have problems due to mineral deficiency or excess.

Those rare occasions which do arise are due to a localized soil deficiency in one element or another, for unless the elements are present in the soil, they cannot, of course, be present in the herbage growing on it. Agricultural advisers and specialist businesses can test the soil and apply dressings if necessary. If a health problem occurs in several members of the herd, or if the herd is being moved into a new location, testing the soil should be considered. Local stock-farmers and their advisers will usually know if a serious deficiency exists. Blood-testing the herd can also be informative, as can forage analyses, but no one of these tests is sufficiently reliable on its own. (See Tables 6 and 7.)

Table 6 Trace elements – ranges reported in soils and lower critical levels in grass

	Co	*Cu*	*Zn*	*I*	*Mn*	*Se*
Range reported in soils (ppm)	5–40	5–100	25–500	2–8	150–3000	0·2–1·25
Critical lower level in grass (mg/kgDM)	0·15	10	40	0·2	40	?

Table 7 Minerals in goat blood – reported ranges in healthy goats

Calcium	2·15–2·81mmol/l	Selenium	282·0–660·0ng/ml
Phosphate (inorg.)	0·73–3·27mmol/l	Copper	9·0–18·73μmol/l
Magnesium	0·83–1·15mmol/l	Cobalt	>300pmol/l
Sodium	141·0–157·0mmol/l		
Potassium	3·5–7·08mmol/l		
Chlorine	102·0–113·0mmol/l		

Vitamins in the diet

There is not very much information yet on the vitamin requirements of goats. Vitamins are traditionally divided into fat-soluble: A, D and E; and water-soluble: the Bs and C. It is A, D and E which are more likely to be deficient in the goat's diet.

Vitamin A (retinol) Goats get their vitamin A from beta-carotene, an orange-coloured substance from green plants, by conversion in the intestinal wall. While beta-carotene circulates in the cow's blood, it does not in the goat's – this is why goats' cream is white. In cattle, beta-carotene has an independent involvement in the maintenance of fertility, but it is not known whether this also happens in goats. Vitamin A is required for the formation of retinal pigments, therefore its deficiency impairs vision. Other symptoms are loss of appetite, unthriftiness, nasal discharge, abnormalities of skin and mucous membranes, lowered resistance to infection, malformed kids and abnormal sperm. The liver contains 75 to 90 per cent of the body's reserves of vitamin A. Excess of this vitamin causes metabolic disorders.

Vitamin D_2 (ergocalciferol) and D_3 (cholecalciferol) Vitamin D is not found in feeds, but is known as 'the sunshine vitamin' because it is made in the skin, induced by ultra-violet rays. It takes part in the absorption, deposition and excretion of calcium and phosphorus. Deficiencies can and do occur in goats kept indoors unless they receive a supplement, kids developing rickets and adults suffering demineralization of the bones (osteomalacia). An excess of vitamin D, on the other hand, causes mineral deposits in the soft tissues.

Vitamin E (alpha-tocopherol) The effects of this vitamin (and there are several substances with the same effect) are closely related to those of selenium, and both are required for normal reproductive efficiency, though some authors dispute this. All are agreed that the most important deficiency disease is white muscle disease (nutrient muscular dystrophy) which can affect the heart or the skeletal muscles. Though naturally occurring deficiencies in adult goats are unknown, kids can be born with white muscle disease in the absence of stored vitamins A, D and E. Colostrum from dams not themselves deficient in these vitamins is very important for new-born kids. Vitamin E is also an antioxidant of fats, and is stored in the body in fat tissue and the liver. It has a relatively low toxicity and upper limits have not been defined.

Vitamin B_1 (thiamine) This and the other B vitamins, also C and K, are produced by the rumen bacteria; as already noted, dietary cobalt is needed for the synthesis of B_{12}. The pre-ruminant kid, however, may benefit from vitamins B_1 and B_{12} in milk replacer. Vitamin B_1 deficiency can occur in kids or adults due to the presence of thiamine-destructive enzymes from eating oak bark, bracken, mouldy feed or over-feeding on cereal grains. The result is the disease cerebrocortical necrosis (CCN, or polioencephalomalacia), which can also be caused by B_1 deficiency due to prolonged diarrhoea or treatment with some drugs, such as some wormers. This fatal disease, causing blindness, headpressing, collapse, etc., can be treated successfully with vitamin B_1 injections if caught sufficiently early.

Figures for the goat's requirements of A, D and E have been estimated from those for sheep and cattle, these estimates being lower for maintenance, with increments for growth and pregnancy and considerable extra allowances for lactation which would take the total daily allowance over those in table 8, which are the latest figures quoted for goats.

Table 8 Daily allowances of vitamins A, D and E for the goat

Vitamin A	3,500 to 11,000 IU	per day
Vitamin D	250 to 1,500 IU	per day
Vitamin E	5 to 100mg	per day

These vitamins are included in mineral/vitamin supplements for adding to goats' feeds. Care must be taken that the expiry date has not gone by before the bag is finished.

Water requirements

Two-thirds of the body is water, and its metabolism in the animal is complicated, yet its importance in the diet is often overlooked. The individual amount needed depends on input and output. Input is a combination of water drunk, water in the food eaten and metabolic water produced within the body. Output is water lost in urine and faeces, water evaporated from the skin and lungs and water lost in milk or foetal development. Normally speaking, input and output must be equal, though each element may vary. For instance, the dry matter and digestibility of the diet will vary, also lactating animals produce moister droppings than dry goats. Urine concentration and volume are controlled by several different mechanisms. Losses by evaporation depend on environmental temperature and respiration rate. Both thirst and the water balance of the body are controlled by the nervous system and hormones. Lactating goats will become dehydrated quite rapidly if denied suitable drinking water. Black Bedouin goats, accustomed to desert conditions, have an extraordinary ability to survive alternate water deprivation and drinking huge amounts, without coming to harm.

If goats are deprived of food, water intake falls; it rises on the other hand with dry food, salt (sodium chloride) in the diet, and high protein feeding. The taste of the water influences drinking, but goats accept bitter, sour, sweet and salty tastes better than sheep or cattle.

Water for maintenance in temperate climates is 1·07g/kg $W^{0.75}$/day.

Water in mid-pregnancy increases to 139g/kg $W^{0.75}$ and this increases again as pregnancy proceeds to about 150g/kg $W^{0.75}$/day. This means that a goat weighing 50kg will drink about 2·8 litres of water per day in late pregnancy.

Water for lactation For each litre of milk produced, a goat must drink about 1·28 litres of water, although a litre of milk itself only contains 0·89 of a litre of water. The remainder is required for the extra metabolic processes involved in producing the milk. It must therefore be remembered that a goat that is a 'gallon milker' must increase her water intake from less than 3 litres/day to over 7½ litres/day from the moment she gives birth.

Extreme environmental temperatures affect water uptake. Over 30 °C, drinking is increased to replace water lost through panting; below 5 °C the amount drunk is reduced.

Goats with ad lib water drink while eating and just after milking; if water is not available at all times, fresh supplies should be given twice daily.

While desert goats can survive and even produce under conditions of water deprivation, Saanen goats will reduce their food intake and their milk yield if they drink only infrequently. Food remains in the rumen for longer, as seen earlier in this chapter, and is digested more fully by bacteria, if drinking is reduced.

Goats are unusual in liking to drink warm water. In very cold weather at least, there is an advantage in humouring them in this. Water must always be clean and uncontaminated, in clean containers. Water from other than the public supply should be tested.

Milk feeding of kids

The oesophageal groove The rumen in the new-born kid is small and not yet functional, so forage cannot be digested. Milk has therefore to be given for at least the first two months of life. This has to be digested in the abomasum; if it were first to enter the rumen, digestive upsets could be caused. Nature has devised a neat trick in baby ruminants to enable liquid milk to by-pass the rumen opening and go straight into the abomasum. This device, the oesophageal groove, effectively closes the rumen entrance. It is formed in response to liquid feeding, by a conditioned reflex, which operates when the kid knows from its experience that it is about to receive a liquid milk feed, either from bottle, teat feeder, bucket or its mother.

If the kid is to be artificially reared, it is important to start it on the feeding method of your choice in the first few days of its life and always stick to the same routine, so that it can see when a milk feed is on its way. If the kid wags its tail and butts with its muzzle when feeding, this is a sign that the oesophageal groove mechanism is in operation and milk is entering the true stomach.

Colostrum This is the thick yellow milk produced by the mother in readiness for the kid's first feeds. It is rich in fat, an energy source which enables the tiny kid to avoid hypothermia. It is also rich in minerals and vitamins, and has laxative properties. Importantly, too, it contains antibodies to the diseases to which the dam is immune and which therefore give the kid vital protection from the pathogenic microbes in the new environment which it enters, totally without existing antibodies, when it is born. These immunoglobulin molecules are large and can only be absorbed through the intestinal wall into the kid's bloodstream during the first day of its life. For all these reasons all kids must have sufficient colostrum feeds starting within minutes of their birth. Colostrum can contain pathogens such as the virus of caprine arthritis encephalitis (CAE) and precautions must be taken to avoid infection of the kids.

Following colostrum feeding, a planned regime of providing adequate goats' milk or replacer must be commenced, leading on to weaning at whatever age is chosen. Further details of this will be given in the next chapter.

Condition scoring

Although the figures given for the nutritional needs of the goat are useful in preventing errors in the diet, there remain many areas of imprecision, not to mention variations in food quality. It is therefore necessary also to check the condition of each goat regularly. This is stockmanship, but the exercise of condition scoring is both a help and a discipline to the stock-person.

It must be stressed that scoring should be carried out, and the results recorded, on a monthly basis, so that a condition profile can be built up over the year. It is the changes in condition during the annual cycle, and any individual departures from the changes in condition seen in the herd in general, which will prove the most help to stockmanship. Practice is needed to acquire the skill. Try to assess goats in very different conditions, while learning.

The scoring method

Holding the goat with one hand, use the other to assess by feel the two regions: the lumbar region (behind the last rib) and the sternal region (under the goat, between and slightly behind the front legs).

The lumbar region: place the hand with the thumb on one side of the backbone and the fingers on the other, and apply pressure to feel the thickness of the long muscles which lie on either side of the upwardly directed vertebral spines, in the angles between these and the transverse processes – the 'eye' muscles. Assess how sharp or rounded the spinous processes themselves feel. Assess how prominent, or fat-covered, the transverse processes feel. Try to put the fingertips under the transverse processes to assess the amount of fat and muscle below them. The diagrams should help to make this more clear. (Fig. 11.)

The sternal region: feel the pad of fat, about 10–15cm long. How thick and how freely movable is it?

There are six grades, from 0 to 5. Experienced graders can give ½ or even ¼ scores.

Grade 0 The animal is emaciated and on the point of death. No subcutaneous tissue can be felt.

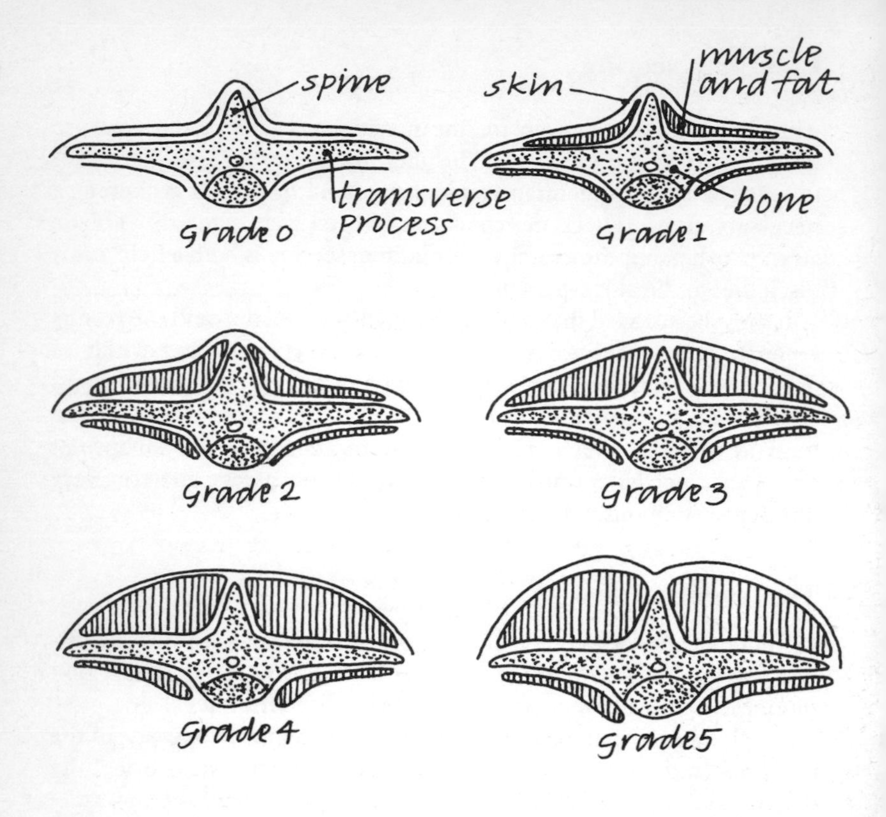

Fig. 11 Condition scoring. Cross-sections of lumbar vertebrae and surrounding tissue.

Grade 1 The animal is emaciated, with ribs, backbone and croup very visible and flanks hollow.

Lumbar: vertebrae rough and prominent and can be held in hand. No muscle or fat. Spines readily felt as separate.

Sternum: pad of fat feels flat and hard, moves easily. Easy to feel the joints and beginning of the ribs with a light touch.

Grade 2 Backbone visible as continuous ridge, croup protuberant.

Lumbar: vertebrae can still be grasped by hand but the transverse processes are now more difficult to feel as there is flesh above and below them. The spines are less prominent.

Sternum: fat pad thicker but still movable. The joints between the sternum and ribs are now covered by a thin tissue layer.

Grade 3 The backbone is no longer prominent and the croup is well covered.

Lumbar: vertebrae thickly covered but can still be grasped with three fingers. A slight hollow can now be felt over the tips of the spines. It is no longer possible to feel the outlines of the transverse processes.

Sternum: the sternal fat is thick and not very mobile. Joints between sternum and ribs can be felt only with difficulty.

Grade 4 Lumbar: the vertebra is too thickly covered in tissue to be grasped and it is difficult to get the fingertips under the transverse processes. The separate spines can no longer be felt.

Sternum: the sternal fat is now almost too thick to grasp.

Grade 5 Lumbar: neither the spines nor the transverse processes can be felt.

The sternal and lumbar scores are averaged for each goat.

Sternum: a separate pad of fat can no longer be identified, because a thick mass of tissue covers the ribs and sternum evenly.

It has been found in France that these two areas give a good indication of the true condition of the goat. A similar system is well used for assessing sheep in this country although goats deposit more of their fat within the body cavity and sheep have more of their fat beneath the skin, so equivalent scores will be higher in sheep. The sternal fat may be difficult to feel in animals under 18 months, and obscured by hard skin in animals of eight years and over. A loss of 0·5 to 1·0 in grade is seen at the end of pregnancy, followed by a further loss of 0·5 to 0·75 points in early lactation. 1·0 to 1·5 points are made up on lush spring/summer herbage, followed by loss of 0·25 of a point in late lactation. A rise in condition around mating times is more likely to enhance fertility than one which stays the same or falls.

In France it has been recommended that, just after drying off, the goat's score should be between 2·25 and 3·5. Just before kidding, it should be between 2·75 and 3·5. After 45 days of lactation, the score should not be less than 2·0 and should not have gone down more than 1·25 since kidding. Thus both very thin and very fat goats require an alteration in husbandry.

In Scotland the advice is that females should have a score of 2·5 at mating. Two months before breeding is anticipated, therefore, the diet of too thin or too fat goats should be adjusted. Many goatlings are far too fat at this stage. The Scottish recommendation that condition should not be less than 2·0 at kidding presumably reflects the somewhat lower condition of hill-grazing stock.

Carcasses are also graded and a score between 2 and 3 is frequently required.

CHAPTER SEVEN

Feeding Practice

The previous chapter established that the goat is by nature an animal evolved to feed on browse – leafy, shrubby material – but sufficiently adaptable to be able to include a certain amount of good grass, and with a great ability to select the best from what is available. In most areas of the highly populated UK, such a diet is impractical, and this chapter is concerned with the practical aspects of feeding the goat in such a way as to maintain productivity and health. It cannot be stated too often that the basis of good goatkeeping is a supply of high-quality hay, and that cereal grains and other concentrated feeds are supplementary to this feeding of hay, and not the other way around! Massive feeding of concentrates can lead to nothing but trouble.

The twelve-month feeding plan

Intensive dairy-goat production Intensive systems are primarily concerned with maximum production at economic cost, unlike extensive systems where the main consideration is utilization of available forage.

During the year, the female dairy goat may have periods of reduced productivity, for instance if she is milking only a little during the first three months of pregnancy. At such times she can be offered lower quality, cheaper feed than during late gestation and early lactation, when her demands for nourishment are at their highest. Daily needs do not always have to be met on a daily basis, however; in early lactation body reserves will be used, even though we must try to minimize their use. Allowance for their replacement must be made in the diet later in the lactation period.

Nutritional needs and condition scoring were set out in the previous chapter. The aim must always be to keep the goats in good condition throughout the year, neither too thin nor too fat. It must also be remembered that intensive production is simply not worth attempting unless the goats concerned are genetically capable of responding to high-level feeding with high productivity. Breeding for improvement should always accompany such an enterprise.

Starting the year about three weeks before mating is anticipated, check the condition score of each eligible female in the herd; it should be 2·5 to 3·0. Those which are too thin should be changed – gradually, as always – on to a diet of ad lib, good quality forage and 1lb (450g) of concentrates per day. Those which are too fat should receive only low-energy, high-fibre forage and a maximum of $\frac{1}{4}$lb (112g) of concentrates per day. It is essential that the condition score be corrected by the end of the third month of pregnancy. This is because during the last two months before kidding, the foetus(es) grow rapidly and require a great deal of nourishment, via their dam. A very thin goat will become emaciated if her very small reserves are drawn upon, while a fat one will be unable, when heavily pregnant, to keep up her feed intake due to the pressure of fat reserves and uterus within her abdomen. Either way, pregnancy toxaemia is the likely result. Damascus goats carrying twins were found to need 16MJ ME/day during the last two months of pregnancy. A good-quality diet should contain, overall, in each kilogram of dry matter, around 11.0MJ ME and 140g of CP.

It is important to maintain as far as possible, by the feeding of top-quality forages, the volume of the rumen during late pregnancy. This has been found to have a great influence on food intake and hence productivity during early lactation. Concentrates can be increased to 600g per day.

The period from kidding to peak milk-yield is the most critical of the year for feeding the high-yielding milker. The diet must be particularly nourishing because the milk-yield reaches its peak before the appetite does; the feed intake is greatest some time between six and ten weeks after kidding, having increased by 30–40 per cent during those weeks. If lack of dietary nourishment forces the too-rapid mobilization of stored energy reserves, acetonaemia (ketosis) is the likely result. Even so, the amount of concentrates fed must not be increased too rapidly after kidding, as the rumen microbe population must be given time to adapt to cereal fermentation. Assuming that, just prior to kidding, plenty of good forage and 1$\frac{1}{3}$lb (600g) of concentrates are being fed, the amount of concentrates can be increased to 2$\frac{1}{4}$lb (1kg) per day for a goat giving a little over a gallon (5 litres) of milk a day, by 3–4 weeks after kidding. This amount must be adjusted depending on the value of the forage fed. If the ration lacks protein, milk-yield will suffer.

As milk-yield falls, the amount of concentrates fed can be reduced, depending on the quality of the hay and other fibrous material also included in the diet. Allowance must be made for replacing the goat's fat reserves, but not to get her too fat. It is no good trying to keep the milk yield up by boosting the feed; the goat 'chooses' the use to which her feed will be put, and the milk-yield will decline. It is suggested in France that 150–400g of concentrate per litre of milk be fed, depending on forage quality, at this stage. The latter figure seems surprisingly high.

Extensive production systems As mentioned above, these systems aim primarily to utilize whatever foodstuffs the environment makes available. As for intensive systems, it is important to use goats which are genetically suited to the method of management. These may be lower yielding dairy, Angora, or cashmere goats, and any breeds being reared for meat production. The aim is to make the most efficient use possible of the land, supplementing the diet when necessary during the year. Use can be made of the ability of the goat to lay down fat reserves in times of plenty and utilize these when there is less natural herbage available. It is important to estimate the value of the plant material being eaten and keep concentrate supplementation to the minimum, otherwise the amount of herbage consumed by the goats will be reduced. The stocking rate – the number of goats per unit area of land – must not be allowed to become too great. As before, the bodily condition of the animals should be regularly assessed.

The effects of diet on milk composition

The milk of goat breeds which yield only sufficient to rear their young is generally high in solids, similar in composition to the milk of sheep. The milk of the high-yielding dairy breeds, on the other hand, has a much lower solids content, akin to that of dairy cows. It is the effect of feeding on the solids content of dairy-goats' milk which gives rise to concern, due to its importance in the human diet, and which will now be considered.

Milk composition, like all biological substances, is variable within certain limits. It can be influenced by breeding, feeding and a number of other environmental factors, but Table 9 gives some typical values. The EC are currently not intending to lay down legally enforceable minimum standards for the solids content of goats' milk, but those who make cheese or yoghurt, or supply milk to processors, will know that

Table 9 Typical milk solids contents of some goat breeds

Breed	*Butterfat %*			*Protein %*			*Lactose %*	*Ash %*
	High	*Average*	*Low*	*High*	*Average*	*Low*		
British Alpine	6·09	4·12	2·94	3·96	2·97	2·5	4·38	0·78
British Saanen	4·87	3·68	2·32	3·61	2·8	2·1	—	—
British Toggenburg	5·13	3·68	2·2	3·38	2·71	2·13	—	—
Golden Guernsey	4·92	4·09	3·23	3·27	2·95	2·54	—	—
Saanen	4·87	3·81	2·71	3·3	2·78	2·37	4·56	0·75
Toggenburg	4·35	3·55	2·21	3·24	2·84	1·82	4·02	0·73
Anglo-Nubian	6·77	5·04	3·66	4·42	3·55	2·7	4·05	0·79
Angora	—	6·0	—	—	4·4	—	4·8	—
Pygmy	—	7·76	—	—	5·06	—	5·35	0·84

milk low in protein and fat results in thin yoghurt and a low yield of cheese, so that the composition of the milk needs to be as rich as possible. Furthermore, those who compete in the British Goat Society's recognized milking competitions receive a disqualification for any goat having less than 3 per cent fat in her milk, and 3·25 and 4 per cent are necessary to qualify for a * or Q* respectively. This regulation has been on the BGS books for a very long time; currently, modern nutritional thinking is likely to result in an award for protein in milk.

The first essential for milk secretion is a plentiful supply of food; fasting rapidly decreases milk-yield. The drop in the amount of milk secreted results in a slight increase in the percentage of protein; the mobilization of the body's fat reserves in an attempt to keep up milk secretion results in a flood of long-chain (C_{18}, see Chapter 11) fatty acids into the system which alters the composition of the milk fat. This becomes richer in C_{18} fatty acids and poorer in C_{10} to C_{16} fatty acids.

Increasing the energy supply is likely to increase the amount of protein produced in the milk through increasing the manufacture of microbial protein in the rumen, but it is easier to influence the fat content of the milk by feeding, and the protein content by selective breeding. As explained in the previous chapter, forage feeding rather than concentrate feeding promotes butterfat production. The type of forage – hay, silage or green forage – does not make a difference. The addition of sodium bicarbonate to a high concentrate diet can help to restore milk butterfat levels, but the magnitude of these differences in butterfat is small – less than 0·5 per cent.

Diets very low in lipids, such as hay, grass silage, some roots, barley and solvent-extracted oil-cakes can result in a reduction in butterfat of up to 0·5 per cent which can be corrected by adding soya-beans, vegetable oil, etc., to the diet, but sometimes this fat supplementation tends to lower the protein content of the milk.

If the diet is deficient in protein, increasing the amount fed may cause the secretion of more milk, but is unlikely to increase the protein content of the milk. Altering the type of protein in the diet has affected only the type of protein, or its amount in the milk, in a few of the experiments carried out. Diets too rich in protein, in relation to energy content, lead to a higher than usual content of urea in the milk. In cows, this reduces the quality of the curd in cheese-making, so it would be wise not to risk it in goats' milk.

To produce a good yield of milk high in solids, a plentiful diet, neither too high nor too low in protein, fat, fibre and digestible carbohydrate must be fed. The digestible organic matter in the diet, which provides the goat with energy, is the most important factor in the secretion of quality milk.

The feeding of youngstock

The importance of feeding first colostrum, then milk, to baby kids was explained in the previous chapter. The regime leading to weaning and beyond is of great importance in producing healthy productive adults.

Feeding on milk or replacer

The first decision to be taken is whether to leave the kids with their dams to suckle, or to take them away and rear them artificially. Successful management of a dairy herd requires knowledge of each goat's milk-yield or, better still, some form of authenticated milk recording; in addition, kids left to suckle a high-producing goat will take a great deal of milk and delay their taking of solid feed and their rumen development. So dairy herd owners frequently separate dam and offspring at four days' post-kidding, when it is safe to milk the udder out. Recording sampling can begin, and the milk will be free of colostrum and therefore usable/saleable for human consumption. At this age, but not much older, it is possible to accustom the kid to teat feeding. Kids should be penned with others of the same age, neither alone nor with others of very different ages and sizes. At first the teat will have to be placed carefully in the kid's mouth (see also Chapter 9). It may be that you will be forced, through disease, to take the kid from the mother at the instant of its birth, and not allow the dam to lick or suckle it – an unhappy situation indeed, and best avoided (see Chapter 10, section on CAE virus).

If Angora, Cashmere, Boer, Bagot or Pygmy goats are being bred, which normally produce enough milk only to rear their kids, the babies will probably be left on their dams. You may have to choose this method with dairy goats too, if you are out all day; in this case you must milk out the surplus milk at least once a day. As soon as the dairy kids are about a month old you can separate them from their mothers at night, if you want the morning's milk for yourself. The difficulty here is that many goats do not 'let down' their milk (a hormonally controlled reflex action) for hand-milking while they are rearing kids, making it impossible to

Plate 8 A Cashmere goat bred in Scotland, adult male. (Photograph: Angus Russel)

strip them out. Goats not stripped out will soon dry off. It is necessary to check that all naturally reared kids are putting on weight as they should: 100–200g per day depending on breed. From a few days old, water, hay and concentrates should be made available. They will take food best at first by copying their mothers, but soon they will have to be fed concentrates away from the dam, who will otherwise eat it all herself. Tiny kids nibble at earth and should be allowed to do so. Some owners prefer the very 'tame' kids resulting from bottle feeding. Though some bottle-fed kids will continue to suckle from their mothers as adults, most kids will be prevented by their mothers from returning to the udder after about two weeks of separation, so that artificially reared kids can graze or exercise with their dams once this stage has been reached.

If artificial rearing is your choice, the next decision is whether to feed goats' milk or 'replacer', which is powdered milk specially made to rear young animals by replacing their mothers' milk. Replacers for kids, calves, lambs and other animals are all available on the market. If you have plenty of goats' milk, there is nothing better; if you can sell goats' milk for more than the cost of replacer, then you will find the latter gives perfectly good results. As with all dietary changes, kids must be accustomed to replacer gradually. If the kid scours (has diarrhoea) you are changing too abruptly, or not following the maker's instructions re dilution, or giving too much, or making it too hot, or too cold, or not taking the strict hygiene precautions that are necessary when feeding all babies – or maybe the kid has an infection. If it seems well despite its loose motions, then correcting the feeding – maybe changing back to goats' milk – may cure it; if it is unwell, seek veterinary help without delay.

Small kids take a little milk frequently from their dams, so 'little and often' must be the best with artificial rearing. Kids drinking from open containers tend to create problems, such as fouling the containers and developing milk sores round their mouths, so most breeders prefer teated equipment, teats made for lambs being suitable. This ranges from individual hand-held bottles in very small herds, through a variety of multi-teated containers, with or without tubes leading from teat to milk source, up to large automatic machines which mix replacer and water as needed, such as that made by British Denkavit Ltd. Suppliers of smallholding equipment can advise and provide what is required.

Milk or mixed replacers can be fed warm (103°F), in measured individual feeds, or cold to be taken ad lib. Automatic machines which feed warm ad-lib replacer may lead to over-feeding and delayed weaning. A typical feeding regime for measured feeds might start with four ½-pint (300ml) feeds per day, spread out as much as possible, increasing, as the kid takes more, to four 1-pint (600ml) feeds a day at about six weeks, then three 1½-pint (850ml) feeds until ten weeks, then reducing the amount and number of feeds to weaning at twelve to fourteen weeks. Early weaning at six to seven weeks has been tried, but may lead to 'weaning shock' and a growth check. While the total solids content of goats' milk is around 12 per cent, milk replacer instructions require a concentration of 15 to 20 per cent. The traditional UK method of giving kids milk until six or even nine months does not seem to be necessary or desirable. Skimmed milk and whey have been successfully fed to kids over one- and two-months-old respectively. In some countries, milk acidified with lactic acid is fed.

If a replacer is to be used, the decision must be made as to which make – there are several on the market. Since cows' milk is very similar to goats', some herd owners use calf replacers to rear kids; these are cheaper as they are manufactured in much greater quantities. In fact a litre of kid replacer may well cost twice as much as a litre of calf replacer, mixed ready to feed. On the other hand, calf replacers contain growth promoters, which most goatkeepers like to avoid as unnatural, also the products made for kids tend to be higher in protein and, importantly, in skimmed-milk powder, rather than whey powder or soya protein. In the kid's stomach, digestive enzymes clot the casein (milk protein) into curds – as when cheese or junket are made – and these clots are gradually digested. This cannot happen with non-casein protein, so digestion has to be abnormal.

The composition of each replacer powder is listed on the bag: skimmed milk powder (SMP), 60 per cent or more in kid products; protein, around 25 per cent – of the dry powder, of course, not when diluted ready to feed; fat/oil about 25 per cent, though Denkavit kid contains only 16 per cent to encourage early weaning; ash, about 7 per cent; fibre nil, indicating a low vegetable content; vitamins and trace minerals. It is up to each breeder to read the labels, follow the manufacturer's instructions for mixing and feeding, and to decide which brand gives the best results in his/her particular herd.

Weaning

When kids of non-dairy breeds are reared on their mothers, weaning takes place naturally, providing arrangements are made to ensure that the kid's share of food is not eaten by the dam, and that precautions are taken to avoid worm infestations if the youngstock graze with adults.

Removal of dairy kids from their mothers, or weaning from artificial rearing, is a more complicated affair. Active steps must be taken to change the young animal's diet from milk or replacer to concentrates and hay because excessive milk feeding, very nice for the kids, will reduce the appetite for solids and delay the growth of the rumen, essential for a future productive herd member. In addition, goats' milk is, presumably, required for human consumption and replacer is expensive in excessive amounts.

Weaning is a particularly big step for a ruminant; the diet changes from one high in lipids and sugars to one high in fibre and starch; the source of energy changes to the products of ruminal fermentation; the anatomy and biochemistry change drastically. For the economic reasons mentioned, early weaning is frequently required in dairy herds;

properly carried out, it does no harm.

Work in France on Alpine kids has shown that though kids can survive weaning – and that means an abrupt cessation of milk feeding – at four weeks of age, this is really too young, and results in a pronounced growth check. Weaning at six or eight weeks was found to be satisfactory. Kids fed on milk, though their needs for energy and protein are being satisfied, will start to eat a little hay and concentrates from four weeks onwards, more so if milk feeding is limited. When weaning is abrupt, it may take six weeks before the kid is eating enough solid food to satisfy its needs. It is important to encourage the eating of good quality forage; if the kid attempts to meet its dietary needs at weaning entirely from concentrates, as it may try to do, then acidosis can easily result.

It may be thought that reducing the milk gradually is better, as recommended in the previous section and as normally practised in the UK. The French have found, however, that this merely reduces growth, and gives a less good result in the end. They recommend that a kid can be weaned if it weighs two-and-a-half times its birth-weight, or if, since starting artificial rearing at four days it has consumed 15¼lb (7kg) dry weight of milk (if female) and 19lb (8·6kg) (if male). It should be eating about 2½oz (70g) of concentrates and 5¼oz (150g) of hay a day before weaning, and not be thin, unwell or suffering from coccidiosis. Even so, early weaning calls for high levels of stockmanship and top quality feedstuffs. The French maintain that 'very late' weaning, at four to five months, gives only marginally better results and is not economic.

Post-weaning feeding

The idea that the large breeds of dairy goat found in the UK need extra time to grow to maturity gave rise to the goatling, and many breeders in this country still do not allow their stock to kid until two years old, although there seems to be adequate evidence that well-fed stock can grow sufficiently large to kid and milk successfully as yearlings. In Cyprus for instance the Damascus goat, not unlike the Anglo-Nubian in stature, is weaned at six weeks and fed a generous diet of 18 per cent protein concentrates and lucerne hay. Almost all these female kids reach 70 per cent of adult weight by seven months, and are mated providing they are not underweight. It is reckoned that this early start adds 17 per cent to the lifetime milk yield and 6 per cent to the weight of weaned kids born to each dam. In France, Alpine kids, fed on hay and concentrates from six weeks, are mated at seven to eight months provided they weigh more than 66lb (30kg). In Norway too, female kids are bred from, providing they have gained at least 3½oz (100g) per day, in order to weigh 77–88lb (35–40kg) (75 per cent of mature weight) after the first kidding. A good weight at this time is necessary for a satisfactory milk yield. A diet sufficiently high in protein is evidently the key to good post-weaning growth in kids, for either dairy replacements or for meat, where intensive management is practised.

Avoiding nutritional disorders

It is obvious that the intake of each food substance – protein, energy foods, minerals, vitamins, etc. – cannot match the body's requirements, minute by minute. Nature has provided an array of mechanisms to cope with this situation. Some substances are excreted, others are stored, others are metabolized into something else and then excreted or stored. The stores can be drawn on in times of need with variable efficiency. The whole system is very intricate and delicately balanced. Sometimes goat owners throw a spanner into the works by placing an intolerable strain on some aspect of the goat's metabolism. Prevention of metabolic diseases is better than cure; when the diseases do occur veterinary treatment is required urgently.

Poisoning

Sometimes the body has no mechanism to deal with a foreign substance: large amounts eaten at one time can cause poisoning (see Chapter 10).

Energy metabolism disorders – ketosis

In late pregnancy, particularly if carrying more than one foetus, and in early lactation, particularly if the milk-yield is high, the goat needs large amounts of energy. The glucose required is made in ruminants, as we have seen, from the fatty acids produced by bacterial fermentation of food in the rumen. If food intake is insufficient, the body's fat reserves (stored energy) are mobilized very rapidly. The biochemistry involved in this causes ketones to accumulate in the blood to a damaging level. One such ketone used to be called acetone, and the disease can be known as acetonaemia. A smell of 'pear-drops' may be noticeable around a badly affected goat. When it occurs before kidding the disease is called pregnancy toxaemia. Hormones may be involved in the onset of disease; growth hormone and probably placental lactogen increase the mobilization of the fat reserves while lack of insulin increases ketone formation. There is evidence that it is lack of glucose, rather than the mobilization of fat reserves, which triggers disease; this coincides with the increased secretion of lactose in the milk-yield, for which the udder requires glucose.

To avoid pregnancy toxaemia, it has been found that the consumption of high quality forage such as lucerne or clover hay should be encouraged, plus about 1lb 5oz (600g) of concentrate per day. Higher concentrate feeding may lead to acidosis and put the goat off her food, which will predispose to toxaemia. A goat which is too fat, and thus has internal fat reserves competing for space in the overcrowded abdomen, is as much at risk as one which is under-fed. Exercise will help to maintain appetite and to prevent ketosis.

Forage feeding during pregnancy helps the goat to eat well in early lactation. At this stage concentrate feeding is permissible to provide the necessary energy, but in order not to put the goat off its food, the amount fed should never be increased by more than 8oz (250g) in a week. Individual goats vary in their likelihood of developing ketosis; it is thought that those with genetically low levels of lactose in the milk may have an advantage. Stress can help to trigger ketosis, and older goats (third lactation upwards) are more at risk.

Mineral metabolism disorders

No matter how much or how little calcium is eaten, the amount passed out in the droppings each day is pretty constant. To maintain a constant level in the blood, the goat has therefore to 'juggle' with the amount absorbed in the intestine, the reserves in the bones, and the amount lost in the urine.

Calcium is absorbed through the intestinal wall bound to a protein, the manufacture of which depends on vitamin D. Milk fever (parturient paresis) is a failure of the juggling act, due to: the large amounts of calcium and phosphorus suddenly required for milk secretion at kidding; the slow increase in intestinal absorption which does not keep up with the increased need; the secretion by the thyroid gland of a hormone called calcitonin which slows down the release of calcium from the bones (opposing the action of parathyroid hormone which speeds it up). Feeding a diet low in calcium in the last weeks of pregnancy will ensure that the goat is actively releasing calcium from the skeleton, which helps to prevent milk fever. If necessary a synthetic form of vitamin D can be given. It is possible that overfat goats may be more prone to milk fever, and that a susceptibility may be inherited.

The treatment of these disorders when they occur is dealt with in Chapter 10.

Foods suitable for goats

Music-hall jokes about goats eating anything are probably based on the facts that they will eat paper, which after all is closely related to their natural foods, and that they feel things with their mouths, rather like human babies. It is true that they will sample a wider range of plants than other farm animals, but they very definitely will not eat anything dirty, contaminated, mouldy, or which has been on the ground and been trodden on, any more than we should. They are also very particular about the smell and taste of their drinking water. As mentioned earlier, they are past masters at selecting the parts of their food which have the highest protein; in fact they very definitely do not 'eat anything'.

Forages fed fresh

Forages, bulk foods, long-fibre foods, as they are variously called, are the most important items of the goat's diet. Because goats select the most nutritious parts and prefer leafy plants to grass, their food is rapidly digested by the rumen microbes and consequently they can eat seemingly enormous quantities of fresh forage. It is impossible (though necessary for dietary calculations!) to know what weight of herbage is consumed by goats while grazing; one recommended menu for a large dairy goat milking heavily allows for 22lb (10kg), fresh weight, of grass to be consumed in a day. Though the labour is very great, goats utilize forages better if these are cut and fed to them in racks than if the animals are free to graze as they wish. A compromise can be to strip-graze the pastures using electric fences. Though goats do not like white clover, they are very fond of many leguminous crops, particularly lucerne (alfalfa). They will eat more (on a dry matter basis) green forage than hay, and more hay than silage, made of the same materials, and more lucerne than rye-grass in each of these three forms.

Lucerne (alfalfa, *Medicago sativa*)

This useful legume is grown on a small scale by domestic goatkeepers (see Chapter 15) and cut to feed green. It can be cut every six weeks, and lasts for about seven years. It should be fed with care to avoid causing frothy bloat. It can be purchased as hay, or dried, chopped and molassed. In all these forms it is highly palatable; it can also be bought compresseed into pellets. Its composition, and those of the other foodstuffs described below, is in the tables at the end of this chapter. It can be seen to be a good source of protein, and of digestible fibre provided it is cut in the early bud stage. Later than this it rapidly loses digestibility through lignification. It is very high in calcium.

Grasses

These vary in palatability and quality, and even the best of them lose their value from flowering onwards; some of the better kinds are: cocksfoot (*Dactylis glomerata*), Italian rye-grass (*Lolium multiflorum*), perennial rye-grass (*Lolium perenne*), various fescues (*Festuca* sp.), meadow foxtail (*Alopecurus pratensis*) meadow grasses (*Poa* sp.) and timothy (*Phleum pratense*). Specialist seedsmen's advice should be obtained on suitable seed mixtures to sow – see Appendix 3 (p. 320). Grasses can be grazed, cut and fed green, made into hay, or purchased as hay or dried grass pellets or, with know-how, equipment and some risk of listeriosis, made into silage. Grasses are higher in calcium than phosphorus and low in magnesium.

Brassicas

Kale and cabbage can be grown or purchased. They are palatable and nourishing but poisonous in large quantities as they can interfere with iodine absorption causing goitre of the thyroid gland, and they can also cause anaemia through destruction of the red blood cells. Feed only after milking, as otherwise the milk will be tainted.

Wild plant food

One reads in books that plant poisoning is rare in goats, but this does not appear to be true, though it is evident that what harms one goat may leave another unscathed. If in doubt, avoid it! Goats will eat a number of poisonous plants, one of the classics being rhododendron (also azalea), which is unusual in causing vomiting among its signs and is often fatal in its effect. There are many published lists of plants which are good food, or alternatively poisonous, for goats. The trouble is that there are many plants which appear on both lists! There is consensus that all evergreens should be avoided. If gathering branches for goats, ash, elm, hazel, willow, hawthorn and elder all seem to be safe. Among the perennial weeds, nettles are outstanding in that their food value is known to be sufficient for it to be worth drying them to make hay. Many other weeds would be dangerous if fed in quantity to stall-fed goats, while a mouthful or two taken while grazing is all right. Foxglove and water dropwort, however, can be fatal in small quantities, while ragwort is infamous for the fact that it is equally harmful fresh or dry. The harmful effects that various plants can have are many and various; books on poisonous plants make interesting, if alarming, reading.

Hay

The importance of good hay in goat feeding has been stressed. What guarantee is there that a particular hay is good? Obviously it must not be mouldy or dusty, it should retain as much of its original colouring as possible and should smell pleasant. But for grass hay these factors do not indicate its nutritive value. Pasture grass has been found to contain almost twice as much protein in April as it does in July, so early-cut hay will have a higher value. This is easy to say, but hay-making is entirely dependent on the weather, both for grass growth, and for drying after cutting; in some very bad years hay cannot be got until August. If grass is going to seed when it is cut for hay it cannot be highly digestible or nourishing. Also, grass will grow on poor soils and can then produce hay poor in minerals. Weather during hay-making is important – rain falling on the drying herbage after cutting washes out soluble nutrients, such as sugars, and reduces palatability. Goats are very fussy about hay, and it is wise to buy a sample bale to try before making a large purchase. Hay containing 'weeds' will be more enjoyed than that made from grasses alone. The selectivity exercised by goats when foraging results in a great deal of wasted hay, rejected and dropped on the floor.

Legume hay, lucerne or giant red clover, are more popular with goats and of greater nutritive value and, alas, more expensive. However, less supplementation with concentrates will be required, so they may prove to be a wise purchase. Those who make all their own hay must beware of repeating any mineral deficiencies in their soil that the goats will encounter when grazing. Most goatkeepers could make small amounts of high quality hay – nettles, for instance – and a strong wooden tripod, with horizontal bars half-way up, is often used for this purpose (see Figure 21, p. 296). The herbage is suspended on the tripod, making sure that there is an air-flow up the middle, until the woodwork is thickly covered. The outer layer may have to be discarded, but underneath should be some excellent hay.

A large, airy building is necessary for storing hay, an open-sided barn being the traditional way. Do not put the bales directly on the floor or they will become damp. Prepare a base of wooden pallets or straw which will allow air to circulate. Rats and mice will be attracted, particularly in winter, and must be eliminated by poison or other means. Do not allow cats to foul the hay, as toxoplasmosis could result.

In recent years, big bales of hay, straw and silage have been produced. While these promote ease of handling to those with tractors and the necessary equipment, they are of no use to small-scale goatkeepers who have only muscle-power.

Dried forages These overcome many of the disadvantages already mentioned for hay, and are more likely to be of an assured quality, though more expensive. Dried grass and dried lucerne can be obtained pelleted, and dried lucerne chopped and molassed. They are conveniently fed as part of the concentrate ration. The heat treatment involved in drying is likely to render the protein more UDP, which is likely to help milk yields. The chopping and grinding, however, theoretically reduce intake, as little or no selection can be exercised, and reduce digestibility as the materials pass through the rumen more quickly. However, on balance, dried lucerne particularly is a very useful product for goats.

Silages

These tend not to be liked by goats as much as hays and fresh forages, and there are dangers of diseases in their use. Nevertheless, silage feeding to goats has increased in recent years, and good silage should be of greater feeding value than poor hay. Some dairy processors will not purchase milk from silage-fed goats.

Silage is forage preserved by microbial fermentation of the soluble sugars it contains. Lactic acid is produced, which pickles the forage, and fermentation stops when the conditions get too acid for the bacteria, at pH4. Air must be excluded and sometimes acid is sprayed on the forage at the start. Plastic bags or sheeting, depending on size and pressure, are used to get rid of the air. If the plastic gets torn, air gets in and the process is spoiled adjacent to the tear. Crops vary in their ensilability; maize and cereal plants are best, then various grasses, then clover and lucerne at the bottom of the list – these have to be wilted to increase the concentration of sugars before being ensiled. Some goatkeepers make silage in dustbin-liners. It is also possible to buy ready-made bags of silage; called 'Goathage' it is wilted lucerne which has been compressed into plastic bags and then sealed.

Things can go wrong in silage making; if a secondary fermentation by clostridial bacteria occurs, the silage will smell bad, be unpalatable and have reduced energy value. If all goes well, however, silage will be of high value providing it was cut when the crop was young – it is usually cut younger than is possible for hay, and loses less nourishment in the fermentation than is lost in hay-making.

Silage feeding can cause acidosis in the goat, unless about 1¼oz (40g)/day of sodium bicarbonate are added to the feed. Listeriosis is another possibility, particularly if the material ensiled was contaminated with earth. The third hazard is mycotoxicosis – rather like pneumonia in goat or man – if the silage is mouldy.

Straws Cereal straws, such as oat straw, are palatable if clean. The leaves are more digestible than the stems, and goats, with their genius for selection, can obtain nourishment from oat straw provided twice as much is offered as they need, in such a way that they can pick it over freely. Pea straw is also of interest, as something different to nibble.

Treatment with alkali increases the digestibility of straw. Caustic soda has been used, but ammonia treatment for two to four weeks is better as it also provides the rumen microbes with nitrogen. It is said that ammonia treatment enables ruminants to live on straw alone, provided enough of it is offered. Full details of the process must be obtained.

By-products Various by-products of industrial processes involved in the production of human food are suitable for goat feeding, and provide a relatively low-cost source of protein and energy. The composition of these materials must be understood so that they can successfully be used as a part of the diet. As explained before, the goat's energy comes from the fermentation in the rumen of both cellulose and starch and sugars. By-products such as molassed sugar beet pulp, maize gluten and brewers' grains are rich in cellulose (digestible fibre). They are likely to be of benefit when the existing diet is short of fibre, either absolutely, or relative to the amount of concentrates being fed. An increase in butterfat, and a reduction in the risk of acidosis, may be expected. Molasses, rich in sugar, can increase the energy content of the diet and enhance milk-yield.

It was also explained that protein may be degraded in the rumen (RDP) or not (UDP). Fish-meal can increase the supply of UDP, necessary for high milk-yields; it also contains oil, calcium and phosphorus.

Depending on the locality, various by-products may be available. Some of value are:

Distillers' dark grains

Palatable, with a high level of energy and protein, at the same time low in starch with a reasonable fibre content.

Maize gluten feed

This can be obtained as pellets or meal; the latter particularly tends to be unpalatable. The protein content is 18–23 per cent, with a very variable digestibility. The oil content can be as high as 55 per cent, which can lead to rancidity.

Molassed sugar beet feed

This is very palatable and consistent, with as much energy as cereals, but derived from digestible fibre and sugars rather than starch. It has 11 per cent protein of which half is UDP, but it is low in phosphorus and needs a mineral/vitamin supplement. It improves milk quality and forage intake and is available as two sizes of nut, and shreds. The home-produced product is better than imported material. It swells up hugely when soaked, so is often fed in this condition. It must not be soaked in a galvanized pail.

Soya-bean meal

One of the best protein supplements, palatable and granular. A full fat product is also available.

Wheat bran

Bran is a by-product of milling wheat. It is frequently used as one ingredient of a mixture of cereals, but would not be palatable alone, if dry. It is traditionally used to make a mash by mixing with boiling water and leaving to stand, when it may be liked by a newly-kidded goat.

There are other by-products, such as fruit pulps – the composition and palatibility of which must be carefully studied if their use is being considered.

Roots

Root and tuber vegetables must really be thought of as concentrates, since they are low in cellulose and high in starch. Since their dry matter is diluted with a lot of water and they need considerable chewing, they are less likely to cause acidosis than cereal grains. As they can be stored for winter use, they can be very useful when no other succulent food is available, but cleaning and chopping them is a chore. Do not cut them into small cubes, which could cause choking. There is less dry matter per unit weight in turnips, swedes and mangolds than there is in carrots, potatoes and fodder beet, and consequently less food value.

It is widely held that male goats must not be given roots, even sugar beet pulp, as to do so runs a risk of blocking the 'water-works' with stones (crystals). It is true, sadly, that male goats, especially castrated ones in which the urethra is particularly narrow, do sometimes die from this blockage, but not that feeding roots is the cause. The offending stones may be formed of oxalate, which comes from the tops of beet not the roots, but they are most often formed of phosphate, when a diet too high in phosphate and too low in calcium, i.e. a diet high in concentrates and low in forage, is fed.

Concentrates: cereal grains and legume seeds

The dangers of cereal feeding have been stressed, nevertheless grains are required to increase the energy density of the productive goat's diet. It is said that crushing and processing are neither necessary nor desirable before feeding cereal grains to goats, since they will crack them for themselves in cudding, and processed grain will be fermented too rapidly. Trials in Cyprus showed that whole grain feeding results in higher butterfat milk than feeding ground and pelleted grain, but oats are frequently fed crushed, and maize, which is very hard, cooked and flaked. There are differences between the nutritive values of various grains, which are reflected in their prices. As a study of the feed composition tables at the end of this chapter will show, flaked maize, which is also particularly palatable, is high in energy, while oats have a high crude-fibre content.

Legume seeds, such as crushed peas, are fed for their higher protein content. They are not, however, as high in protein as the by-products of oil extraction, such as soya meal.

Minerals

It is essential to supplement the diet of goats with extra minerals. Blocks, which are licked by the animals whenever they feel the need, should always be available. Largely salt (sodium chloride), they also contain, typically, magnesium, iodine, selenium, zinc, iron, cobalt and manganese in appropriate quantities; one suitable for goats rather than, say, horses should be purchased. Two sizes are available, one about the dimensions of a building brick, for which wall-mounted holders are available; the other, larger, a cube with a hole through the middle. The latter is very heavy and care must be taken that it cannot fall on to a small kid. Lick-bricks should be placed under cover from rain.

A further supplement, rather than being consumed at the discretion of the goat, is a powder placed in the concentrate mixture. If cereals, etc., are bought in the form of a ready-made 'goat-mix' a mineral powder will almost certainly have been added, though the label may not be more explicit than to say 'minerals'. Vitamins may have been added too – it would be wise to enquire about this. Otherwise, if making up a mixture of cereals, protein supplements, and so on, at home, one is responsible for adding one of the many commercially available mineral and vitamin powders. It is safer to add the appropriate dose at the time of feeding, as if a large quantity is mixed under home conditions, the mineral powder will tend to fall to the bottom of the bin resulting in a very uneven supply to the goats. Mineral powders are made for different species – those made especially for goats tending to be more expensive – and in standard, 'high phosphate', 'high magnesium', etc., varieties. Unless problems are encountered, a standard mineral/vitamin mix is probably best. Typically this will contain calcium, phosphorus, sodium, magnesium, selenium, iodine, cobalt, manganese, zinc, and vitamins A, D and E – an expiry date for the vitamin content will be printed on the bag. The usual dose is 25g/day.

Concentrate mixtures

Several feed manufacturers make mixtures for goats. The contents are either just a loose mixture (coarse ration) or the ingredients have been ground and pressed into pellets, cubes or nuts. The coarse ration gives the goat the opportunity to reject any ingredient it does not fancy, while pellets do not. If you make up your own feeds, you know what they contain; if you buy ready-mixed ones you do not, though there is a move towards a more full declaration of the contents on the label. Due to the apparent relationship with cattle disease, it is currently illegal to feed any animal-derived material to ruminants. Whether or not the ingredients are listed on the label, you will find a declaration of the protein content (probably 16 or 18 per cent), and the oil, fibre, ash and vitamin contents. This does, however, leave you wondering about the Metabolizable Energy (ME) content, which is given only by a minority of manufacturers, and enquiries on this score would have to be made. It is noticeable that 'information' leaflets provided with these mixes vary wildly in the amount that they recommend should be fed, so one should make up one's own mind on this important matter.

Planning the menu

Formulae and calculations are very off-putting to most of us, but it is worth the struggle to have guide-lines on how much, and what, ought to keep our goats healthy and productive, without spending much more than we anticipate. Goats vary tremendously in size, and their weight has a big influence on how much they need to eat. Most owners underestimate the weights of their stock, so if you do not have suitable weighing equipment it is worth buying a goat weigh-band which will enable you to estimate the goat's weight from its girth.

This done, we need to consider what foods we can obtain or grow, what storage space we have, what foods will be palatable and at the same time value for money, and what the goat is producing or how fast it is growing.

Figures for the goat's requirements were given in the previous chapter; tables of the composition of some suitable foods are at the end of this one. The menu is composed by putting the two together, remembering that at least half the dry matter of the diet should be in the form of forage, and that the goat can only eat a certain amount in one day.

How much food (dry weight) in one day?

It was explained in the last chapter that work in France has shown that many factors influence food intake; as a working rule, however, we say that a dry goat will eat 3·5 per cent of its body weight per day, while one milking heavily will eat 5 per cent of her own weight. Growing stock, and lower-yielding milkers, will put away 4–4·5 per cent per day.

Table 10 Maximum food intakes (dry weight) per day

	Daily milk yield (kg)						
Weight of goat (kg)	*0*	*1*	*2*	*3*	*4*	*5*	*6*
	Food dry matter intakes (kg/day)						
10	0·45						
20	1·1						
30	1·3						
40	1·4						
50	1·5	1·7	1·9	2·1	2·3	2·4	2·5
60	1·8	2·0	2·2	2·4	2·6	2·8	3·0
70	2·1	2·3	2·5	2·7	2·9	3·1	3·3
80	2·4	2·6	2·8	3·1	3·4	3·7	4·0

How much energy and protein must the daily intake contain?

Let us consider a housed dry goat first, requiring maintenance alone. It was stated in Chapter 6 that maintenance energy and protein requirements are:

0·445MJ ME and 2·5gDCP/kg $W^{0.75}$/day.

A good-sized female dairy goat, weighing 70kg, has a metabolic weight of $70^{0.75}$kg, which is 24kg. She therefore requires

24 × 0·445MJ ME and 24 × 2·5gDCP per day, which is
10.68MJ ME and 60gDCP per day,

and she must obtain these from eating 2·1kg (Table 10) dry weight of food per day.

Dividing 10·68 by 2·1 gives us the 'energy density', in MJ ME/kg, of the foods she will find sufficient, and this value is exactly what we find in the appropriate column of the food composition tables. We can do the same for the protein:

10·68 divided by 2·1 is 5·1MJ ME/kg and 60 divided by 2·1 is 28·5g DCP/kg.

The tables indicate that even untreated barley straw would provide these amounts. Supposing we did decide to feed our unfortunate goat entirely on barley straw, we must consider her mineral requirements. Chapter 6 shows that maintenance requirements of calcium, phosphorus and magnesium are 0·067, 0·046 and 0·017g/kg of liveweight respectively. So a 70kg goat needs 4·69g, 3·22g and 1·19g respectively. Inspection of the composition tables indicates that barley straw is too high in calcium and too low in phosphorus. A small cereal grain supplement would be one way of putting this right.

The requirements of dry goats are very, very much less than those of goats giving milk. The previous chapter pointed out that for every litre of 4 per cent butterfat milk produced per day, the goat needs an extra 4·95MJ ME per day, and 50g of DCP for each litre of 3 per cent protein milk. Table 11 gives the energy and protein requirements of different weights of dairy goat giving different yields of milk, with this level of fat and protein. A 70kg goat giving 6kg of milk per day needs 40·4MJ ME and 360gDCP. Her increased appetite will allow her to consume 3·3kg dry weight of food per day, so she will need food with an 'energy density' of 40·4 divided by 3·3 which is 12·2MJ ME/kg. The food tables plainly show that this will not be met by forage alone.

Table 11 Energy and protein requirements for lactation

	Milk-yield (kg/day, 4% fat, 3% protein)													
	0		*1*		*2*		*3*		*4*		*5*		*6*	
Liveweight	*ME*	*DCP*	*ME*	*DCP*	*ME*	*DCP*	*ME*	*DCP*	*ME*	*DCP*	*ME*	*DCP*	*ME*	*DCP*
(kg)	*MJ*	*g*	*MJ*	*g*	*MJ*	*g*	*MJ*	*g*	*MJ*	*g*	*MJ*	*g*	*MJ*	*g*
50	8·3	47	13·3	97	18·2	147	23·2	197	28·1	247	33·1	297	38·0	347
60	9·5	53·8	14·5	103·8	19·4	153·8	24·4	203·8	29·3	253·8	34·3	303·8	39·2	353·8
70	10·7	60	15·7	110	20·6	160	25·6	210	30·5	260	35·5	310	40·4	360
80	11·8	66·8	16·8	116·8	21·7	166·8	26·7	216·8	31·6	266·8	36·6	316·8	41·5	366·8

How much concentrate supplement is needed?

We are feeding grass hay ad lib, estimated from the tables to have 8·8MJ ME/kg. The 'goat mix' says on the label that it gives 13·2MJ ME/kg.

If we call the amount of concentrates we are to feed 'c', then

$$12{\cdot}2 = 13{\cdot}2c + 8{\cdot}8\,(1 - c),\ \text{so}\ c = 0{\cdot}77$$

The goat is able to eat 3·3kg/day of dry matter, and we have found out that the dry matter content of the goat-mix is 90 per cent (i.e. 0·9kg/kg, the rest being moisture), so the amount we give the goat per day is

$$\frac{0{\cdot}77 \times 3{\cdot}3}{0{\cdot}9} = 2{\cdot}8\text{kg}$$

This represents 2·5kg dry matter in a total of 3·3kg, 75 per cent of the ration as concentrates, which is much too high for the goat's health. Sugar beet pulp, which is high in digestible fibre, cellulose, could come to the rescue. A possible diet could be as follows:

Fresh weight (kg)	*Dry weight (kg)*	*ME in amount fed*	*DCP in amount fed (g)*
Hay: 1·2	1·0	8·8MJ	59·0
Sugar beet pulp: 0·6	0·5	6·3MJ	38·5
Goat mix: 2·0, at (18% CP)	1·8	23·8MJ	259·0, at (80% of CP)
Totals	3·3	38·9MJ	356·5

Thus the 70kg goat giving 6kg of milk a day will get almost enough energy and protein from a diet of 0·6kg of sugar beet pulp and 2kg of goat mix with a 13·2ME and 18 per cent crude protein, provided she is fed hay in sufficient excess so that she actually consumes 1·2kg. There is still only 31 per cent of the dry matter of her diet being provided as forage, however, which is less than one would like to see.

Using the figures given for requirements under various circumstances in the previous chapter, the sort of calculating methods shown above, and the food composition tables which follow, one can set up suitable diets for the herd, varying the ingredients.

There remains the question of value for money of any items we may consider feeding. It is not the price per kilogram that matters but the unit prices of energy as ME and protein as DCP. The method is to find the unit prices of ME and DCP in a mixture of two standard ingredients of known price, then using the technique of simultaneous equations we can make a comparison with the item we are considering.

Take barley at £110/tonne and soya meal at £150/tonne. From feed tables, barley has 12·9MJ ME/kg dry matter, and 82g/DCP/kg. The dry matter is 86 per cent of the fresh weight, so on a fresh weight basis, barley has 11·1MJ ME/kg and 70·5 DCP/kg. Similarly soya meal has 11·1MJ ME/kg and 407g DCP/kg.

Let x be the unit value of ME(£/MJ) and let y be the unit value of DCP(£/g DCP)
then $11{\cdot}1x + 70{\cdot}5y = 110$ (price of barley)
and $11{\cdot}1x + 407y = 150$ (price of soya)
so $x = 9{\cdot}15$ and $y = 0{\cdot}12$.
So the unit price of ME is £9.15/MJ and the unit price of DCP is £0.12/g, in the barley and soya-meal mixture.

The item we are considering is, say, brewers' grains, which has a dry-matter content of only 22 per cent and this dry weight contains ME of 10·4MJ/kg and DCP of 154g/kg (from the tables). The comparative worth of brewers' grains is: $0{\cdot}22((9{\cdot}15 \times 10{\cdot}4) + (0{\cdot}12 \times 154))$. This equals 25, so brewers' grains is not worth feeding instead of the barley/soya-meal mixture if it costs more than £25 per tonne.

The information given in Table 12 is taken from the MAFF 'UK Tables of Feed Composition and Nutritive Value for Ruminants', 2nd edition, 1992, and other sources. The abbreviations used are:

ME	Metabolizable energy in MJ/kg DM
DOMD	Digestible organic matter in the dry matter in g/kg DM (D value)
Dig CP	Digestibility coefficient of crude protein
DM	Dry matter as organic dry matter in g/kg fresh weight
CP	Crude protein in g/kg DM
EE	Ether extract in g/kg DM (i.e. lipids)
CF	Crude fibre in g/kg DM
Ca	Calcium in g/kg DM
P	Phosphorus in g/kg DM
Mg	Magnesium in g/kg DM
UDP	Undegradable protein in g/kg DM

A dash indicates that the information is not available.

Table 12 Nutritive values and composition of foodstuffs

	ME	*DOMD*	*Dig. CP coeff.*	*DM*	*CP*	*EE*	*CF*	*Ca*	*P*	*Mg*	*UDP*
Hays											
Good quality grass	10·1	631	0·6	866	114	13	291	5·0	2·5	1·5	20
Poor quality grass	7·4	519	0·44	859	84	14	366	5·0	2·5	1·5	—
Average lucerne	8·5	583	0·75	865	183	13	302	15·6	3·1	1·7	36
Good red clover	8·9	—	0·65	850	161	—	287	—	—	—	32
High-temperature dried crops											
Dried grass	10·4	653	0·82	894	188	37	213	7·4	3·3	1·8	—
Dried lucerne	8·8	544	0·67	895	199	28	—	15·0	3·0	2·3	—
Straws											
Barley	6·4	450	0·19	867	42	14	394	4·2	1·1	0·7	8
Oats	7·2	496	0·07	846	34	14	394	3·9	0·9	0·9	7
Wheat	6·1	429	0·23	872	39	12	426	3·9	0·8	0·9	5
Oilseed rape	5·5	416	0·64	865	62	19	—	19·8	1·2	1·1	—
Ammonia-treated oats	8·0	550	0·26	843	75	18	—	4·6	1·4	1·1	—
Fresh forages											
Cabbage	13·7	781	0·87	107	207	17	182	8·3	1·9	1·5	82
Young grass	12·6	765	0·77	193	190	25	130	5·9	3·0	1·1	76
Mature grass	7·5	564	0·41	204	97	16	288	5·9	3·0	1·1	—
Kale, thousand head	11·9	765	0·85	143	188	20	200	12·5	4·2	1·6	75
Lucerne	9·4	600	—	220	205	—	282	17·0	3·0	—	—

Table 12 Nutritive values and composition of foodstuffs (*continued*)

	ME	*DOMD*	*Dig. CP coeff.*	*DM*	*CP*	*EE*	*CF*	*Ca*	*P*	*Mg*	*UDP*
Silage											
Grass	10·9	678	0·67	255	168	43	300	6·4	3·2	1·7	34
Lucerne	8·0	557	0·68	338	194	25	296	17·6	3·0	1·8	77
Energy feeds, roots											
Fodder beet	11·9	829	0·51	183	63	3	53	2·8	1·8	1·6	—
Potatoes	13·4	882	—	204	108	2	38	0·4	2·0	1·0	—
Swedes	14·0	876	—	105	91	4	100	3·5	2·6	1·1	22
Carrots	12·8	—	—	130	92	—	108	—	—	—	—
Turnips	11·2	—	—	90	122	—	111	—	—	—	24
Energy feeds, cereals											
Barley	13·3	838	0·77	864	129	16	53	0·9	4·0	1·2	26
Maize	13·8	879	—	873	102	39	24	0·1	3·0	1·3	39
Oats	12·1	740	0·73	858	108	41	121	0·9	3·4	3·0	22
Wheat	13·7	889	0·77	857	128	17	26	0·6	3·3	1·1	25
Energy feeds, by-products & fruit											
Apples	11·9	856	0·24	136	38	12	—	0·5	0·9	0·3	—
Apple pomace	9·1	608	0·24	242	69	27	184	1·6	1·4	0·6	—
Citrus pulp, dried	12·6	825	0·56	890	68	22	—	14·6	1·1	1·7	—
Maize gluten	12·9	765	0·77	885	220	44	39	2·3	9·3	4·1	44
Sugar beet pulp, dried molassed	12·5	823	0·71	876	110	4	203	7·6	0·8	1·1	—
Wheat bran	10·8	698	—	892	174	39	114	1·1	12·6	6·2	34

Molasses	12·7	—	—	750	41	—	0	—	—	—	—
Whey	14·5	—	—	66	106	—	0	—	—	—	—
Protein supplements											
Field beans	13·1	800	0·82	848	267	13	85	1·3	8·6	1·9	53
Brewers' grains, wet	11·5	589	0·77	250	218	62	186	3·5	5·1	1·7	—
Brewers' grains, dry	10·3	600	—	900	204	—	169	3·3	4·1	—	82
Distillers' dark grains, barley	12·2	612	—	907	275	67	—	1·7	9·6	3·3	—
Distillers' dark grains, maize	14·7	726	0·78	889	317	110	—	1·4	8·4	3·2	—
Herring meal	16·4	805	0·96	913	758	95	—	34·4	23·6	2·1	—
White fish-meal	13·4	737	0·93	914	716	80	0	57·2	33·2	2·0	430
Linseed meal	11·9	850	—	885	391	87	102	3·4	8·7	5·4	156
Peas, field	13·5	909	0·89	866	254	14	63	1·0	5·8	1·6	105
Soya-bean meal	13·3	844	—	886	493	18	58	3·9	7·4	3·0	104

A note on seaweed meals

The analysis of any brand of seaweed meal varies with the types of seaweed composing it and the time and place of harvesting. Three main types of seaweed are used: knobbed wrack (the most commonly used), tangle and bladderwrack. A full typical analysis of each is given over, along with the advertised analysis of a 'Natural Mineral Mixture', based on seaweed meal, but utterly different from them in mineral balance. References to seaweed meal in this book do not include such mixtures.

Composition per cent	*Knobbed wrack (Ascophyllum)*	*Tangle (Laminaria)*	*Bladderwrack (Fucus)*	*'Natural Mineral Mixture'*
Digestible protein	6·82	6·1	0	9·82
Carbohydrates	58·09	45·8	53·6	56·08
Fibre	4·5	8·6	9·4	4·2
Ether extract	2·26	1·1	4·2	2·4
Ash	18·85	16·8	16·3	16·82
Moisture	9·48	16·3	11·3	10·68
Sand	1·48	?	?	1·02
Calcium	1·49	4·62	1·60	1·29
Phosphorus	0·12	0·21	0·11	0·82 (n.b)
Magnesium	0·71	?	?	1·30
Sodium	1·97	?	?	1·17
Potassium	2·50	1·25	1·80	2·87
Chlorine	1·11	0·69	1·32	0·91
Iodine	0·07	?	?	0·16
Cobalt	trace	trace	trace	0·11

CHAPTER EIGHT

Selection of Breeding Stock

This chapter is intended to serve the needs of the person who wishes to produce goats' milk in quantities, great or small, and not those of the pedigree breeder whose primary purpose is to produce stock for sale. Consequently the science of genetics is only referred to in the broadest terms, and the available space is devoted to the selection and breeding management of stock suited to specific methods of goat dairying.

The dairy farmer's breeding policy is to provide a regular flow of milk, and an adequate supply of satisfactory youngstock for herd replacement with the minimum of cost and risk. In selecting his breeding stock he must always play safe, and resist the lure of gambling in genetics. In particular he must avoid as far as possible the dangers of inbreeding, and at the same time say farewell to his hopes of achieving perfection in his herd. For the pedigree breeder, a knowledge of genetics is useful; and an intimate knowledge of the strains of the breed in which he is working is essential. UK goat breeds are of such recent origin that the only breed characteristics which are established with any degree of uniformity are those affecting colour and appearance; inbreeding is therefore essential to the pedigree breeder who wishes to produce a strain of distinctive character and true-breeding potentialities. The pedigree breeder must not only take the risk of inbreeding to fix desired character, but must proceed to the even more expensive business of proving and publicizing the character produced; he must design his feeding and management to ensure the maximum possible production from his goats, have their yields officially recorded and their confirmation officially approved in the show ring. Maximum possible production is never the most economic production; the pedigree breeder's overheads will usually ensure that the milk of his herd is expensive to produce. One of the main handicaps to the development of a goat-dairying industry in Britain is that far too many goat owners accept the expensive responsibility of the pedigree breeder, and far too many pedigree breeders lack the stock or personal qualities to justify their status. There is a dearth of first-rate goat-dairy farmers, and a surplus of second-rate stock breeders.

The main considerations in selecting stock for milk production are their inherent milking capacity, food-intake capacity and food-to-milk conversion efficiency. The pedigree breeder generally aims to produce an animal with all these qualities developed as highly as possible. But the degree to which the dairy farmer needs each of these distinct qualities will depend upon the system under which his goats are managed. While it is not possible to discuss the special needs of every conceivable method of management, useful distinction can be made between three main types of management: i.e. *stall feeding* or *yarding*, under which system the goats have most of their food cut and carried to them; *goat farming on improved land*, in which case the goats obtain the bulk of their ration grazing crops which have been specially grown for them and other farm stock; *free range on scrub and rough grazings*, in which case the goats obtain the bulk of their ration foraging growth which would otherwise be wasted. Each of these systems of management calls for a different set of qualities in the goat who is to make the most of them.

The labour of feeding the stall-fed goat is generally the biggest item on the cost sheet, with the price of the food running it a close second. The type of goat needed is one which will produce the maximum of milk for the minimum of food and attendance. The goat's milking capacity should be high, and her efficiency in converting food to milk of the very best; but there is no call here for a big food-intake capacity. Some modern dairy goats have been developed to consume a prodigious daily ration of cheap roughage and convert it into milk. But roughage which is dirt-cheap on the ground where it grows becomes progressively more expensive every time it is handled; and the dearest foodstuff, per pound of digestible protein, fed to farm stock.

One solution to the economic problem set by the appetite of the modern dairy goat is to feed it large quantities, not of roughages but of highly digestible succulents – cultivated fodder crops – and a generous measure of concentrates. This system is as good as any for forcing

Plate 9 SM CH. §§44/36†Mostyn Minimayson BR. CH. Sire of seven full champions and many other prominent prize winners and milkers. Six times winner of Stud Goat Cup. So indelibly has he stamped his beautiful quality and conformation on his daughters and passed on the great milking ability of his forebears, that breeders from far and wide have travelled their goats for his services. Minimayson must unquestionably be termed 'The Greatest'. S. Breaston Byrnian BR. CH. D. RM44 Mostyn Minimay Q*1.

yields to the maximum for official recording, and is popular with pedigree breeders. For the domestic goatkeeper with a few goats, who can grow most of the special fodder crops in the garden in his or her spare time, it is also suitable. But the high cost of field cultivation and handling of bulky and sappy crops is a handicap to commercial dairying.

In countries such as Italy and Spain, where commercial dairying with stall-fed goats is widely practised, the type of goat used is a small, flat-sided, milky creature such as the Maltese or Malaga goat (which are probably derived from the wild *Capra prisca*). These goats are fed on industrial by-products such as tomato-cannery waste and olive pulp, plus a little hay and a small concentrate ration. They consume little more than half as much food in proportion to body weight as our leading breeds, but produce almost as much milk in proportion to body weight as our 200-gallon (909 litres) milkers. The *Capra prisca* derivatives seem adjusted to a more concentrated ration than is healthy for goats derived from the mountain goat *Capra aegagrus*. In Italy and Spain they are preferred to breeds with bigger food capacity and higher yields per head for village milk supplies and small enterprises; higher yielding British Saanens were imported by Spain from England for a large specialist goat dairying enterprise.

There are several British industries, at present pouring their edible wastes into rivers, pig troughs and sewage pipes, which might support stall-fed goat dairies. Though they do not appear in the show ring or the leading herds, there are a number of flat-sided goats with small food capacity and good milking abilities. The true *Capra prisca* type, with a short, straight, twisted horn, are rare.

Whether we adopt the fodder crop or industrial waste as a basis of feeding for stall-fed or yarded goats, we are adopting a very *wet* diet; this is an essential to maximum production from a given quantity of nutrients. The goat which will make the best use of such a diet is not the one who has the biggest capacity for bulk, but the one with maximum capacity for slop – that is, a goat that is deep but not wide in the body.

It is also interesting to note that the goats used for commercial dairying on the stall-feeding system in the Mediterranean basin are long-haired. On the other hand, the free-range foragers, even in mountainous districts, are preferred short-coated, like the Murcian and Granada goats of Spain. Goats derive their heat mainly from the bacterial fermentation of roughage, and on a concentrated diet are more subject to chills, even when housed, than when full of fibrous forage on the bleakest mountain top. Our stall-fed goats are often insulated from the cold by flannel coats or a layer of subcutaneous fat; flannel coats are less foolproof and fat is more expensive than long hair, which can always be clipped away from the udder for the sake of dairy hygiene.

The goat farmer on improved land has need of quite another kind of goat. His labour costs are largely concerned in the organization of controlled grazing; his food costs are moderate. The fewer goats he has to control, the better; the more forage crop they will eat, the less they will need out of the bag of expensive concentrates. He needs goats with as high a milking potential as possible, and a good food-intake capacity. A very high food-to-milk conversion efficiency is not always desirable; fodder crops are succulent rather than fibrous, and the sudden chills of the British climate are notorious. Goats grazing improved land will resist chilling and feed in worse weather if they carry a little fat or a coat of long hair (the flannel substitute is unsatisfactory out of doors, especially in the rain). The weight of a big udder, a well-filled and capacious paunch and a rather fat body are most easily carried on short, sturdy legs. Ground clearance for the udder is of little account in cultivated fields, but the udder requires to be exceptionally well hung, broad in the base and close to the body to avoid the dangers of chilling.

The leading breeders produce excellent stock for this purpose.

Goats that have free range on rough grazings are consuming food that costs next to nothing; the more they eat of it and the more efficiently they convert it to milk the better. Big food capacity is of prime importance. Such goats consume ample fibre to keep themselves warm and have no need of a quilt of fat or long hair, so their food-to-milk conversion efficiency may be very high. But very great milking capacity is not wanted. Any advantage to be gained from a superlative yield is more than cancelled out by the large proportion of concentrate necessary to maintain it, the burden of carrying an unwieldy udder over rough ground, and the damage that udder will suffer from the brambles and hazards of a typical rough grazing. Long legs are needed to give the udder ground clearance, and are an important asset to a forager among trees and shrubs. The quality and hang of the udder is not a matter of primary importance. As a free-range goat farmer, the writer has happiest memories of some tough, pendulous bags, with skin like crêpe rubber, impervious alike to knocks, scratches and climate. The classic peach-skinned vessel, delectable to handle and easy on the eye, is so firmly attached that it is the loser in every encounter; after a day on range its tenderness may make more difficulties for the milker than a

Fig. 12 Conformation of scrubland and mountain goats. (a) The Murcian goat has the scrubland shape; (b) the Toggenburg goat has the mountain shape.

more unsightly and tougher bag.

The amount of damage that the udder of a heavy-milking goat receives, while on the range among scrub and rock, depends to some extent on the general conformation of the goat. This matter has received less attention than it deserves from our breeders. Viewed from the side, most of our goats show an almost horizontal top line, from the shoulders to the rump; if anything the shoulders are a little higher than the top of the rump; the under line slopes steeply down from the chest to the udder. This conformation gives the goat, in appearance and reality, perfect balance and control on steep slopes; it is typical of mountain goats the world over, and has been impressed on our stock by the influence of the Swiss breeds. Goats adapted to scrubland, as opposed to mountain, pastures have a different shape; the top line slopes quite steeply up from the shoulders to the top of the rump; and the under line is roughly horizontal from the top of the udder to the chest. This conformation gives better ground clearance to the udder; the weight of the udder is more equitably shared between fore and hind legs; and the extra-long hind legs give the goat a higher reach when browsing. It is exemplified in the heavy-milking Murcian goat of Spain, in the little scrub goats of Galicia and in many of the Oriental breeds. Of home-made breeds, the Anglo-Nubian possesses this character, a legacy from its desert ancestors, and the British Alpine shows the character to a rather smaller extent.

Where goats have a very extensive range, it is a matter of importance that their herdsman should be able to see them at a distance. The camouflage markings of the Toggenburg and British Alpine are a negation of utility in this respect, and the nondescript colouring of the Anglo-Nubian is little better. Pure white or unbroken black are the colours most easily spotted at a distance. Those who doubt the visibility of black should ask a hill shepherd the colour of his dog.

There are eight recognized dairy breeds of goat in Britain.*

(1) The Saanen

Nominally of Swiss origin, goats of this breed are all descended from imported stock, the majority of which came from the flat fields of Holland. White in colour, placid in disposition, rather short in the leg and capable of the very highest yields, this breed is well suited to goat farming on improved land. The labour involved in avoiding unsightly staining of the white coat is a disadvantage if the breed is kept under intensive conditions, where the goats spend much of their time lying in their own droppings and urine. White-skinned Saanens develop skin cancer in sunny climates; dark-skinned, but white-haired, goats are immune. Short legs and big udders associate badly with the brambles and hazards of a rough grazing. Udders in this breed are usually shapely and well hung. Butterfat percentages are often good – around 4 per cent.

(2) The British Saanen

Of mixed origins, this is usually a considerably heavier and slightly leggier goat than the pure Saanen, but otherwise similar. Udders are less shapely on the whole; but yields as high as any. Like the Saanen, and for similar reasons, it is best adapted to goat farming on improved land, where the tendency of some strains to run to fat does not go amiss.

(3) The Toggenburg

Descended from imported Swiss stock, this is numerically the weakest of the recognized breeds; and progress with the breed in this country is hampered by lack of numbers and inbreeding. Brown and white in

*Plus the 'British', which are registered crosses.

Plate 10 Miss Mostyn Owen with a prize-winning group. CH:RM49 Mostyn Daphne Q*8 BR.CH., lifetime yield 38,516 lb – a British record; CH:R40 Mostyn Marjenka Q*1; Mostyn Marypoppin, goatling daughter of Marjenka, subsequently CH:RM46 Q*1.

colour, with an active but affectionate disposition, the Toggenburg is a small goat, sometimes under 100lb (45·4kg) adult weight. Yields have improved in recent years, some now reaching 220 gallons (1,000 litres) per annum, though butterfats remain low, averaging around 3·5 per cent. Udders are usually well hung. Not every household requires 1,000 litres of milk a year. As a stall-fed household goat, the Toggenburg has the advantage of requiring less food than most, and its brown coat resists staining. As a free-range rough grazer the Toggenburg is in its traditional environment, and on a lower level of nutrition the butterfat content of its milk is not so outstandingly bad. There is a good demand for pedigree stock for export.

Plate 11 The first-ever Toggenburg champion, R142 Spean Meliflous Q*BR. CH., bred and owned by Mrs J. Shields. Champion at nine shows in 1979.

(4) The British Toggenburg

Of mixed origin, this breed carries a large proportion of Toggenburg blood, which is needed to fix the rather elusive Toggenburg colour and markings. But it is a big 63kg goat, giving notably high yields of milk with a butterfat percentage of around 3·6. The high metabolic rate necessary for heavy production, superimposed on the warm-hearted Toggenburg disposition, tend to make the breed remarkably excitable. Though this may be a slight disadvantage in a strictly commercial herd, it makes the 'BT' the most responsive pet in the goat world. The relatively stain-proof coat is an added advantage under intensive conditions. Long legs, high-geared temperament and big food capacity suit the breed for free range on rough grazings, but badly hung udders may occur. As the BT is restrained with more difficulty than other breeds, and its rather low butterfat is further lowered by soft, rich feeding, it is not the obvious choice for goat dairying on improved land.

(5) The British Alpine

This shares much ancestry with the British Toggenburg, but is black with white Swiss markings; a big, leggy goat, turning the scale around 140lb (64kg), it is capable of high yields, with a butterfat percentage of about 4 per cent. It is the only British-made breed to have been developed mainly under free-range conditions, and has an independence of character which would be better appreciated if there were more men farming goats, and fewer women keeping them as pets. Under intensive conditions, the BA is almost as easy to keep clean and as efficient a producer as the British Toggenburg, but a duller companion. For arable dairying, it is rather less easy to control than the Saanen or British Saanen, and is relatively free from the tendency to run to fat; but unwieldy udders are not uncommon in the breed, and give rise to trouble under these conditions. The black coat attracts flies, which is a handicap in closely wooded country. It remains the most adaptable all-round breed, and could be developed into a first-rate goat for scrub dairying with less difficulty than any other.

(6) The Anglo-Nubian

Of mixed origin, this breed owes its distinctive features to imported goats of Indian Jumna Pari and Egyptian Zaraibi type. Undoubtedly the most important British contribution to world goat breeding, the Anglo-Nubian is the most distinctive of our recognized dairy breeds in both appearance and performance. Colours are various, with roan and white predominating; lop ears and Roman nose are typical. It is more heavily fleshed than the Swiss breeds, weighing about 150lb (68kg), gives yields comparable with the Toggenburg in quantity (up to 300 gallons [1,364 litres] a year), but with a butterfat percentage of around 5 and a percentage of solids-not-fat of 10 to 11, compared with the 8 to 9 of other breeds. The rich milk characteristics of the flat-sided goats of the sub-tropics are here blended with the great fodder capacity of the Swiss breeds to make an economic producer of what is perhaps the most highly digestible and perfectly balanced food available to mankind.

The udder of the Anglo-Nubian, though not remarkable for its shapeliness, has a better ground clearance than that prevailing in other breeds. Its coat is short and lacks under-down. Goats do not lay down subcutaneous fat like sheep, and this goat does not withstand adverse weather like the longer-coated breeds. It is excellently suited to free-range goat farming, or to arable dairying, under which conditions it can produce milk *solids* as economically as any. But being an inefficient producer of milk *water*, unless an exceptionally discerning human consumption market will pay a premium for its quality, Anglo-Nubian milk is best cashed when made into cheese, yoghurt or fed to stock. As a house goat the Anglo-Nubian has the disadvantage of being exceptionally vocal; but its milk is undoubtedly the most desirable of any. Due to its prolificacy and rapid growth it makes an excellent meat breed.

(7) The Golden Guernsey

A survival from the Channel Islands, which owes its existence to Miss Miriam Milbourne and her painstaking breeding programme. The first of these goats were imported from Guernsey in 1965 and two years later the English Golden Guernsey Goat Club was founded. British Goat Society recognition of the breed followed in 1970 with the opening of a Golden Guernsey Register. An English Guernsey section followed in 1974, building on the first cross of GG female to S, or BS male. In 1989 a new regime for breeding-up English Guernseys came into force, based on the use of GG or EG males, and there are Supplementary Register (G), Foundation Book (G) and Herd Book (G) registers in the *BGS Herd Book*, which are operated in the same manner as for the other dairy breeds. Golden Guernseys are listed by the Rare Breed Survival

Trust due to the lack of separate bloodlines, but are fast approaching the threshold limit for this listing. There are now two breed societies, one in Guernsey, the other in mainland Britain. Like the Toggenburg, the milk-yield is a more manageable quantity for family use, averaging 218 gallons (990 litres) in a year, with good solids content.

Ideally, a Golden Guernsey should have erect ears, slightly curled at thc tip, straight or slightly dished facial outline and be horned, hornless or disbudded. Its colour should be golden as should its skin, and its coat should be short, with or without long hair on spine and quarters. It should be light-boned, of slender build, with fairly long legs, straight back, and ribs deep and well sprung, although larger goats than this ideal are acceptable if they are not coarse. The neck should be slender without tassels, but variations from the ideal include neck tassels, long coat and shades of gold in coat colour from pale honey cream to deep gold.

(8) English

Bred to be hardy, with a winter coat containing plentiful under-down, and to produce milk for a long lactation period on an economical diet, the English goat must present an appearance of soundness. Neck tassels, that hallmark of the Swiss goat, must be absent. The colour is frequently brown, with a black stripe down the back and black legs. Type is very important in this attempt to re-create our original native goat and, prior to registration, every goat is inspected by officers of the English Goat Breeders' Association. The milk-yield is intentionally family-sized, rather than commercial. Currently this breed is not registered by the British Goat Society.

Plate 12 Novington Boris GG 85H. Winner of many prizes at open shows including Best Adult Male and Reserve Best Exhibit, Devon Goat Society Show; Best A.O.V. Male, Warwickshire Goat Society Show; 3rd Sire and Progeny, Cornwall Male and Young Stock Show. Breeder Miss R. E. Carney, owner Mrs M. Rosenberg.

Table 13 Relative popularity of BGS registered breeds in Britain, 1945–90 (Numbers and percentages of registration in the *BGS Herd Book*)

		1945	*1950*	*1960*	*1970*	*1974*	*1978*	*1986*	*1990*
Anglo-Nubian	No.	58	118	103	224	454	870	1,346	939
	%	(1·9)	(4·0)	(9·9)	(12·9)	(14·7)	(10·9)	(17·2)	(18·5)
British Saanen	No.	683	587	180	329	470	1,187	1,367	880
	%	(22·4)	(20·1)	(17·3)	(19·0)	(15·3)	(14·9)	(17·4)	(17·3)
Saanen	No.	132	108	58	70	94	156	220	206
	%	(4·3)	(3·7)	(5·6)	(4·0)	(3·1)	(1·9)	(2·8)	(4·1)
British Toggenburg	No.	154	169	148	246	343	829	1,086	652
	%	(5·1)	(5·8)	(14·2)	(14·2)	(11·1)	(10·5)	(13·9)	(12.8)
Toggenburg	No.	29	30	24	32	46	101	184	157
	%	(1·0)	(1·0)	(2·3)	(1·8)	(1·5)	(1·3)	(2·4)	(3·1)
British Alpine	No.	376	316	99	128	197	356	434	411
	%	(12·3)	(10·8)	(9·5)	(7·4)	(6·4)	(4·5)	(5·5)	(8·1)

British	No.	640	737	274	393	686	1,519	1,633	900
	%	(21·0)	(25·3)	(26·3)	(22·7)	(22·3)	(19·2)	(20·9)	(17·7)
Foundation Book	No.	365	310	52	92	262	854	416	213
	%	(12·0)	(10·6)	(5·0)	(5·3)	(8·5)	(10·8)	(5·3)	(4·2)
Supplementary Register	No.	610	541	102	219	504	1,661	638	213
	%	(20·0)	(18·6)	(9·8)	(12·6)	(16·4)	(20·9)	(8·2)	(4·2)
Identification Register	No.	—	—	—	—	—	325	228	94
	%	—	—	—	—	—	(4·1)	(2·9)	(1.8)
Golden Guernsey	No.	—	—	—	—	23	—	274	355
	%	—	—	—	—	(0·7)	—	(3·5)	(7·0)
S.R. (G)	No.	—	—	—	—	—	1	2	53
	%	—	—	—	—	—	(0·12)	(0.02)	(1.0)
Total		3,047	2,916	1,040	1,733	3,079	7,924	7,828	5,073

Selecting a goat

If you want great milk capacity, look for a long lean head, interested eyes, and lively movements, a long, slim neck and gently sloping rump; feel for a fine, pliable skin with a smooth and lustrous coat, for a large, elastic udder and knobbly milk veins; if possible, consult the milk records for the goat or her dam and sire's dam.

The detection of these basic utility qualities in the genetic make-up of the male goat is not a suitable task for the eye and hand of inspection – not even if the eye and hand be those of an expert. In selecting a herd sire, his progeny are the only reliable guide to his quality; his pedigree may give grounds for reasonable hope.

While his skin quality, character and conformation have a similar significance to the parallel qualities in the female, their appearance in the male goat may be easily confounded by secondary sexual characteristics and accidents of rearing.

The operation of the scent glands on the he-goat's skin, and his habit of spraying his front legs and neck with urine occasionally set up in the

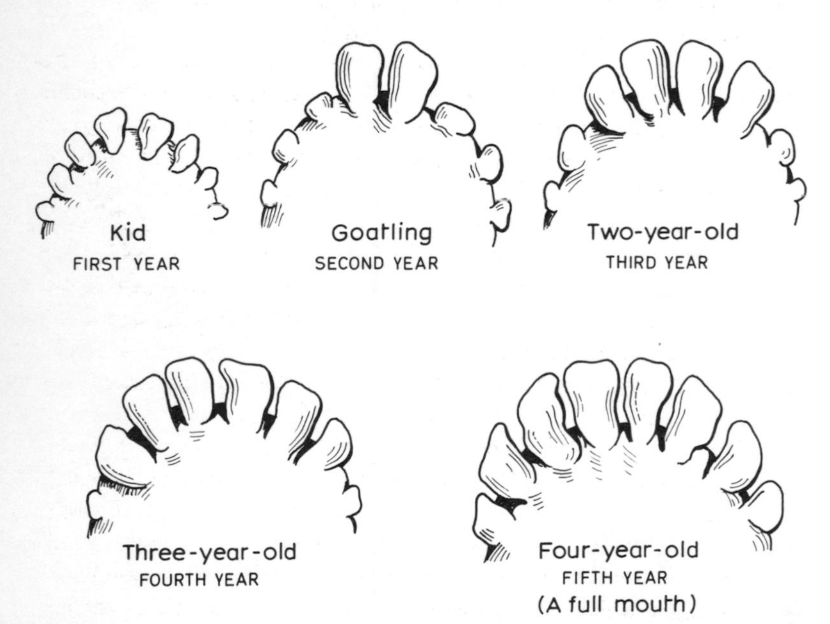

Fig. 13 Determining the age of goats using the front teeth of the lower jaw.

housed billy a kind of dermatitis which converts a naturally soft and pliable skin into an unpleasant hairless corrugated leather. The practice of feeding male kids large quantities of milk until they are far past the natural weaning age is liable to result in a distorted growth – great length of bone and great depth due to the sag of the milk-filled stomachs, but far less width and spread of rib than that to which his inheritance entitles him. The need under common circumstances to keep the males in a small enclosure and carry their fodder to them may lead to them receiving a rather concentrated ration, which accentuates the appearance of narrowness without in any way affecting what inherited capacity they may have for breeding wide-ribbed kids.

There are also certain characteristics that may be required of the herd sire, but are not particularly wanted in milkers. Exceptional docility, for example, is in common demand. A well-exercised he-goat accustomed to kindly but firm control, is only dangerous when roused. But if chained and penned for life and handled with some fear and disgust, any but the poorest in spirit becomes unmanageable. Though a very soft-tempered male seldom breeds outstanding milkers, or a temperament suited to free-range life, poor-spirited and effeminate billies are justifiably popular with those who lack the facilities for keeping a more vigorous beast under control.

The odour of billy is universally unpopular, a menace to the production of palatable milk, and an all-round nuisance. Its seasonal production is a secondary sexual characteristic of the goats of cool regions, in which there is a limited breeding season. In tropical countries, where the male remains sexually potent all year round, scentless males are not uncommon and the odour of the odoriferous is milder. The tropical ancestry of the Anglo-Nubian has resulted in some strains of the breed producing very mildly scented males and there seems to be no theoretical reason why this highly desirable characteristic should not be further propagated. Effeminate males of other breeds are often mildly scented too; but in the Swiss breeds, such males frequently prove unsatisfactory breeders.

Contamination of the clothing of the billy's attendant is unavoidable, and the regular attendant becomes inured and insensitive to the smell after some months. But whenever contaminated clothing comes within the range of the public's nostrils, damage is done to the goats'-milk industry. It is an elementary precaution to handle the billy and milk the goats in distinctively different overalls. A radical solution of the problem is offered near the end of Chapter 9 (p. 184).

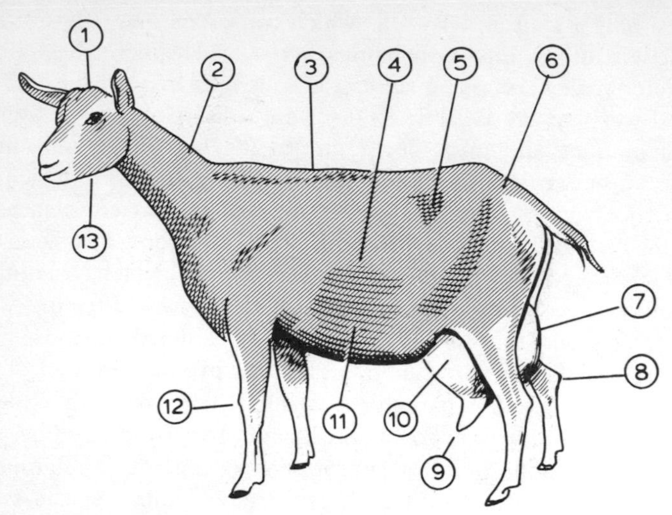

Fig. 14 Points of the productive milking goat

(1) Long, lean head with lively expression.
(2) Long, lean, silky-skinned neck.
(3) Strong, straight, muscular back.
(4) Deep, wide-sprung ribs, last rib curving back.
(5) Hollow in front of hips bespeaks capacity of digestive organs.
(6) Long, gently sloping rump to support heavy udder.
(7) Capacious udder, broadly based on the body and with fine, elastic skin, swells the rear profile.
(8) Hocks are sufficiently straight to avoid bruising udder when the goat walks.
(9) Teats are hand-sized and distinct from the rest of the udder; the thicker they are, the quicker she will be to milk.
(10) Extension of udder forward is the goat's safeguard against udder chills.
(11) Under the belly you can feel, if not always see, the big, knobbly, milk veins.
(12) Strong, clean bone in front legs.
(13) Wide, powerful jaw to match a big appetite.

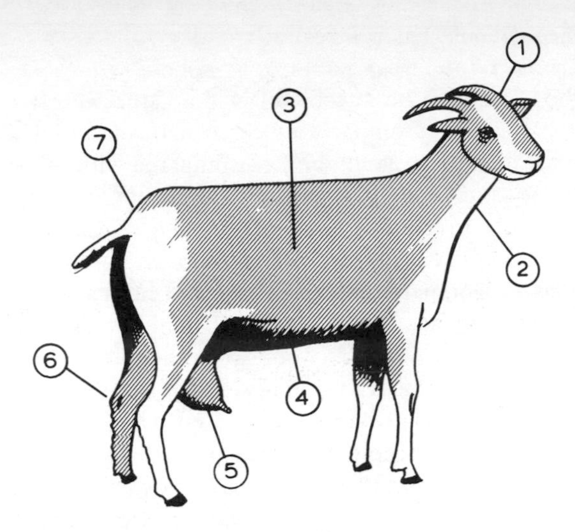

Fig. 15 Points of the unproductive milking goat

(1) Short, convex face, and more or less upturned nose.
(2) Short, coarse neck.
(3) Shallow, straight ribs, and brick-shaped profile.
(4) Small belly.
(5) Small, tough-skinned udder, purse-shaped with little, finger-breadth teats.
(6) Hocks nearly together as she walks.
(7) Short, steep rump.

She is expensive as a present to 'a good home', unless the she goes straight into the pot.

Progeny testing As mentioned earlier, male goats likely to upgrade the productivity of the herd cannot be selected by eye, but must be chosen for their genetic superiority. For dairy goats, nothing can be done unless milk-yields are properly recorded. But this is only the start; high-yielding progeny may be so because of genuine genetic superiority, or because of lavish feeding and management. The need to eliminate environmental effects does not seem to be fully appreciated among many goatkeepers. Lack of centralized sire selection and breeding plans is resulting in the continuation of the sorry state of affairs that, even among those goats whose milk-yields are publicly recorded, only 25 per cent reach 220 gallons (1,000 litres) in 365 days; and yet a few reach over 550 gallons (2,500 litres), so it can be done!

The method found to be successful in the cattle world is that of Improved Contemporary Comparisons. Sires on test are used in several different herds, then their daughters are compared with others in the same herd at the same time. First lactations over a three-month period are compared, and corrected for seasonal differences. Each male must produce at least twenty daughters, spread over a number of herds, and their lactations be compared with the national herd's performance. Sires whose daughters are better than their herd contemporaries are given a plus figure; those who are worse get a minus figure.

There are several problems standing in the way of adopting this method in the dairy-goat world in the UK. High-yielding goats are not required by all, and those who do want them usually have small herds, so that many breeders would have to co-operate in any scheme. Those who have large herds where a sire could father several daughters are usually in business and registration fees are expensive when the profits have to be carefully watched. Nevertheless co-operation between breeders appears to be the only way in which genetic gain in the national herd could be achieved.

Artificial insemination and embryo transfer These tools of the breeder's trade have been developed in recent years so that a high degree of success can now be assured. The British Goat Society licenses those wishing to carry out these procedures in dairy goats, and has altered its Regulations to cover the registration of kids obtained by these methods. The British Angora Goat Society performs the same services for Angora goat breeders, and no doubt the British Cashmere Goat Society will follow suit, particularly as artificial breeding has been of particular benefit in developing more highly productive cashmere goats in this country. The organization Caprine Ovine Breeding Services Ltd (COBS), affiliated to the British Goat Society, holds stocks of frozen semen and trains extractors and inseminators. The EC have produced rules to ensure the health and hygiene of trade in genetic material, but operators in the UK are already working to high standards in this respect; Codes of Practice are available, including those published by the BGS and BAGS. Of course, only genetic material of the highest value merits artificial breeding techniques, which highlights once more the need for evaluation.

Breeding schemes These too, are only possible if breeders will agree to co-operate. A breeding scheme for Cashmere goats is currently in operation in Scotland. The object is not only to identify superior animals, but to spread their superiority through the herds belonging to the scheme as quickly as possible. The plan is to establish one central élite herd and a small number of satellite herds. A typical scheme would be to pick two good sires (this is easier for fleece-bearing goats where the male is productive as well as the female) and, housing them on the élite farm, to collect and freeze their semen. Half the females on each satellite farm are inseminated, the other half are served by the farm's own males. Male progeny are then evaluated on one particular day, the best two selected – and so on. Sometimes the best females from the satellite farms are taken to the élite farm to speed up genetic gain. Gradually the productivity of all the farms in the scheme increases.

Genetic nomenclature for goats Those who study the genetic make-up of goats need to know what they are talking about, right round the world. A committee was established to standardize the names of the known genes in goats (and sheep). Workshops were held and reports are published from time to time, the first being produced in 1986 (see Bibliography).

CHAPTER NINE

Breeding Problems

The demands that mankind makes of the goat have not required any fundamental alteration of its breeding mechanism, nor distortion of its shape and proportions. In this respect the goat is a great deal more fortunate than its comrades of the farmyard, and consequently remains relatively free from breeding troubles.

The naturally early maturity and high fecundity of the goat is adequate to meet any normal demand for the maintenance or increase of stock. The problem of winter milk supply is solved by allowing the goat to 'run through' alternate years, and run a twenty-two-month lactation as a matter of course, in domestic herds. Commercial herds, however, need to increase their milk production in winter by kidding some of the herd in the autumn. This can be done by manipulating the day length, or by using hormonal treatment.

Goats in temperate climates are seasonal breeders, the length of the breeding season becoming shorter as distance from the equator increases. Natural mating occurs, in temperate zones, when the days are shortening; supposedly this results in the kids' being born, five months later, into a supportive spring environment, but in the UK, births any time between December and August are common; late kids seem to show a reluctance to grow as rapidly as early ones.

Shortening day-length is perceived by the eyes, which are connected by a complex linkage of nerves to the pineal gland, close to the brain. This secretes melatonin during darkness; increasing night length, through increased melatonin secretion, is one of the key factors in starting the breeding season, though temperature, nutrition and the sudden introduction of a male can also play a part. From April to July, females do not come naturally into season, i.e. they are anoestrous.

Males also have a breeding season, the rut. During the autumn and winter months they produce increased amounts of more fertile semen, and the secondary sexual characteristics, referred to in Chapter 8, increase.

The female cycle

The hypothalamus (floor of the fore-brain) produces two releasing hormones. The first of these, Follicle Stimulating Hormone-Releasing Hormone (FSH–RH) causes the anterior pituitary gland, just beneath the brain, to release Follicle Stimulating Hormone (FSH), a substance that stimulates the ovaries, at each tip of the horns of the uterus, to develop a blister, inside which one of the store of eggs in the ovary rapidly develops. This blister itself secretes oestrogen, a substance that produces the symptoms of oestrus (heat). The uterus contracts, the cervix at the mouth of the uterus relaxes and opens, the vagina is tensed and lubricated by the discharge of mucus. The goat becomes restless, bleating and wagging its tail, with a red and swollen vulva often showing some of the discharge. Vagina and uterus are prepared for the entrance of the male sperms, and the goat will 'stand' to the billy. Oestrogen also has a negative feed-back effect on the pituitary gland thus checking over-production of FSH-RH, and it triggers the secretion of the second releasing hormone, Luteinizing Hormone-Releasing Hormone (LH-RH).

When the blister in the ovary reaches its full size, the pituitary produces LH (luteinizing hormone), causing the blister to burst and the mature egg to start on its passage down the long, twisted Fallopian tube which joins the ovary to one horn of the uterus. In the site of the egg-cell, the corpus luteum develops, which secretes progesterone, which has the opposite effect to that of oestrogen. The outward symptoms of 'heat' subside, the vagina relaxes and dries, the cervix closes to seal the uterus, and the uterus relaxes and is richly supplied with blood. If service has been effected, the egg on its passage down the Fallopian tubes encounters a sperm and is fertilized; arriving in the uterus it finds the place prepared for it by the action of progesterone, implants and develops. The corpus luteum remains, continuing to secrete progesterone until the foetus is mature.

Progesterone also inhibits the secretion of FSH-RH, and thus the cycle is delicately balanced, with ovulation usually occurring late in oestrus. In a non-pregnant goat the uterus secretes, after prolonged progesterone stimulation, prostaglandin, which, after about a fortnight, destroys the corpora lutea. Progesterone secretion therefore stops, removing the negative feed-back and allowing the cycle to start all over again through renewed secretion of FSH-RH.

The main function of the pituitary in this matter is to ensure that kids are born only if and when conditions are suitable for their survival. Under tropical conditions, where seasonal changes of climate and vegetation are immaterial, the goat's breeding season extends over the twelve months of the year; lactations are short where Nature provides abundant food for early weaning, and two or more pregnancies a year are normal. Normal health is the only stimulus needed to maintain the cycle. Farther from the Equator, seasons become more pronounced. Generally speaking, in the Northern Hemisphere, to which some goats are native, the spring months of April and May offer the kid the most auspicious welcome. As the goat, like the sheep, carries its young for five months, the months of October and November are ideal for starting the breeding season. Various factors signify the arrival of these months: diminishing length of day, falling temperatures, herbage growing drier and tougher. Each of these factors plays its part in stimulating the onset of the breeding season. In Kenya, where the length of day does not change greatly, but there are two periods of rain each year, followed by lush growth and drought, the hardening of the herbage with the drought starts the breeding season, and the kids are born after the rains; so there are two breeding seasons a year. Farther north, and throughout Europe, the dominant factor in the onset of the breeding season is the diminishing length of day.

Our goats are derived from a wild stock native to the Mediterranean basin, and their glandular control is still designed to produce kids in April and May in these latitudes. The onset of the breeding season in southern France and the Balkans is in mid-October, and the kids arrive in April. But the critical rate of change in the length of the day takes place closer to midsummer the nearer we go towards the North Pole. In the extreme north of Scotland the length of the day is diminishing as rapidly at the end of July as it is in mid-October in Marseilles. Consequently, the goat's breeding season in the north of Scotland starts at a time of year which results in the production of kids in the first days of January, and a very cold welcome they get.

For three weeks after mating, the fertilized egg lies free in the uterus, nourished by a secretion of the uterine glands. After three weeks the outer, trophoblast, layer of the developing egg eats into the mother's tissues and forms an attachment – the placenta. For the next ten weeks the trophoblast cells extract all the needs of the growing embryo from the mother's tissues, whether she can spare them or not. Level of feeding of the mother during this period has no effect whatever on the growth of the embryo, which extends its attachment to cover the area of uterus wall available. Then eight weeks before kidding, the trophoblast cells die off, and the foetus is fed by transport from its mother's bloodstream into its own. The quantity of diffusion depends on the area of the placental attachment, which will be less than the optimum when the goat is too small, or too fat, or carrying too many kids for her size. The quality of the exchange the foetus receives will depend on the current diet of the mother and her metabolizable body reserves.

Parturition in the goat is initiated by the unborn kid itself. The foetal pituitary secretes adrenocortiocotrophic hormone (ACTH) which causes the foetal adrenal glands to secrete cortisol. (This is why cortisone can cause abortion). This boosts the synthesis of placental oestrogen, which increases prostaglandin secretion by the placenta, which causes regression of the corpora lutea and hence a sharp fall in

Table 14 Goat Gestation

Mated in	*Will kid on* (*Mating date less no. below*)	
August	January	−3
September	February	−3
October	March	−1 or −2
November	April	−1 or −2
December	May	−1 or −2
January	June	−1 or −2
February	July	0 or −1
March	August	−3
April	September	−3
May	October	−3
June	November	−3
July	December	−3

This table is based on an average gestation period of 150 days, and the date a goat may be expected to kid is calculated by subtracting from the mating date the number indicated in the table. A goat mated on 10 September, for instance, would be expected to kid about 7 February, whereas a goat mated on 17 April should kid about 14 September. Where February falls during the 150 days, allowance is made for leap years.

plasma progesterone level. This last is what initiates the birth; contractions of the uterine wall begin. The increased oestrogen:progesterone ratio causes the maternal posterior pituitary gland to release oxytocin, which has an effect on uterine contractions and cervical dilation.

During pregnancy, as mentioned above, luteinizing hormone from the pituitary gland maintains the corpus luteum, which secretes progesterone. The placenta secretes caprine placental lactogen, which also has a luteotrophic function. During the last 15 days of pregnancy, as foetal cortisol levels rise in the dam's bloodstream, placental lactogen secretion falls, also promoting regression of the corpora lutea.

The end-product of the sexual cycle is not only the kid, but lactation. A number of different hormones control the development of the udder and the secretion of milk; progesterone from the ovaries, prolactin from the pituitary gland and placental lactogen. As each foetus has a placenta, this hormone is produced in greater quantities the more kids the dam carries, a neat ploy of nature to ensure more milk for more babies.

Goats may show a rise in milk-yield just before coming in season, then a fall during the heat period.

The udder may become very full before the kids are born, but do not start to milk the goat unless the udder becomes really hard with over-distension, as doing so will remove colostrum which the kids need.

The male goat

The male sex organs are a little more complicated than outward appearance suggests. The testicles in which the sperms are produced consist of a number of minute tubes, converging into a central tube. The sperms are formed in the small tubes, move into the central tube and thence into the epididymis, a very long convoluted tube in which the sperms mature. From the epididymis they emerge into the cord connecting the epidiymis with the seminal vesicle above the bladder, where the sperms are diluted in the seminal fluid which activates them – the seminal fluid being produced by a number of contributory glands.

The pituitary of the male goat, like that of the female, produces both FSH and LH. LH stimulates the testicles to produce both sperms and the hormone *testosterone*, which gives the billy his typically male appearance and behaviour. A deficient secretion of LH is symptomized by a feminine appearance, sluggish service and low fertility, and is a condition which may be inherited or result from faulty feeding.

Hornlessness and infertility

A hornless or 'polled' mutant turns up from time to time in all the horned species – cattle, deer, sheep, goats, etc. It is a tentative measure of disarmament, a saving of expense on defensive weapons; provided the defensive weapons have ceased to serve a purpose, the mutation is likely to prove successful. But in animal societies such as those of the red deer and the mountain goat, in which the males select the pastures for the herd, and for whom the main natural enemy is starvation, though female horns be irrelevant, the annual renewal of the stag's antlers and the ever-growing burden of the male goat's horns are society's sole guarantee that the male's choice of pasture matches the needs of the mothers. Although the stag does not lead his family party as the male goat does, it is the summer parties of antler-growing stags that pioneer new territory for the expanding red-deer herd. For both deer and goats, a hornless male may be social disaster. Inevitably the hornless male will keep in better condition, and be more mobile, than his horned or antlered rival. During the breeding season his advantage must ensure him a disproportionate share of services; if fully fertile, he must reproduce his kind, to the destruction of the flock.

Under natural selection the proportion of hornless males must not exceed about 5 per cent, and increased use of hornless males should boost herd fertility.

The hornless mutation that has survived among domesticated goats in Europe almost certainly originated among the mountain goats of Switzerland; indeed all the investigations of the factor, in Israel, Britain, Germany and Japan, have been carried out on descendants of exported Swiss Saanens. This mutation meets the ecological requirements of the mountain-goat flock in the following way:

The factor for hornlessness is inherited as a Mendelian dominant: if the gene is present, there are no horns, if absent, there are normal horns; there is no half-way stage. But hornlessness is closely linked with the factor which modifies fertility. In heterozygous form in females it causes an increase of 5 per cent in fertility by increasing the incidence of twinning and triplets; in homozygous forms in females it causes a partial, or apparently complete, pre-natal sex change; no true homozygous females are born, but a mixture of pseudo-females, pseudohermaphrodites, and pseudo-males; all, of course, hornless, and all infertile. In the homozygous form this factor is known to affect the potency of males. The infertility of the males shows itself in two distinct forms: either under-sized testicles, or a blockage in the seminal ducts in both testicles, causing the accumulation of sperm to form a small abscess. Few or no viable sperms are produced in either case.

The practical implications of this state of affairs are perfectly straightforward. There is no possibility of establishing a truly hornless breed of goats, as no true-breeding hornless females are born. (To be exact, possibly one in a thousand hornless females is true-breeding, but she goes unrecognized.)

But it is still worth propagating the hornless factor, for convenience and for economy; and for the slightly greater fecundity, where that is desired. None of the progeny of a horned/hornless mating will be sexually abnormal, and half or more should be hornless. In a hornless/hornless mating, if the male is heterozygous, one in four of the female embryos will become intersexes or pseudo-males; if the male is homozygous, half the female embryos will be affected.

The loss caused by intersexes arises, not so much from the visibly abnormal kid, which is destroyed at birth, but from the pseudo-male or pseudo-female, which is reared to maturity before its worthlessness is discovered. So it is seldom worth rearing the apparent female offspring of a homozygous male and a hornless female; the chances of normality are no more than evens. To exclude homozygous hornless males, it is necessary in practice to exclude all hornless males, both of whose parents were hornless, until they have sired a horned kid from a horned female, and so proved their heterozygous state. Rearing the hornless offspring of the mating of a hornless female with a heterozygous hornless male is still not a commercial proposition: setting the cost of rearing against the selling price of the in-kid goat or buckling, the loss on rearing one intersex or pseudo-male eliminates the profit on several normal kids.

Interestingly, it seems that occasionally two horned parents do produce a polled kid. Amateur research, published in the British Goat Society's *Journal* for February 1990, indicates that this happens in certain families.

Fig. 16 (a), (b), (c) summarizes the situation in conventional symbols.

The horned goat is *homozygous* – that is, the factor for horns is inherited from both parents. The factor for horns, being recessive, would be apparently suppressed if a factor for hornlessness were inherited from either parent. Linked to the factor for horns is a dominant factor for normal sexual development. The horned goat passes on to all its offspring a 'recessive' factor for horns and a 'dominant' factor for normal sexual development.

The homozygous hornless goat inherits from both parents a factor for hornlessness which is dominant, linked with a factor which tends to change the foetus in the uterus into a male. If the foetus is genetically male the intersex condition may be associated with small testes or a granuloma at the head of the epididymis. If female it will be born apparently female, or obviously intersex – but always sterile. The sex-change factor is recessive, effective only in the homozygous state. There are no homozygous hornless females capable of breeding, but the males pass on the dominant hornlessness and recessive sex-change factor to all offspring. Homozygous males may have small testicles and reduced fertility.

The heterozygous hornless goat inherits from one parent the recessive factor for horns, linked with the dominant factor for normal sexual development, and from the other parent the dominant factor for hornlessness with the recessive factor for foetal sex-change. The goat appears hornless, and its sexual development is normal; but it can pass to its offspring a factor either for horns or hornlessness, each with its linked factor for normal or abnormal sexual development. All hornless females capable of breeding are heterozygous.

Fig. 16 Hornlessness and the intersex condition (a) The three genotypes.

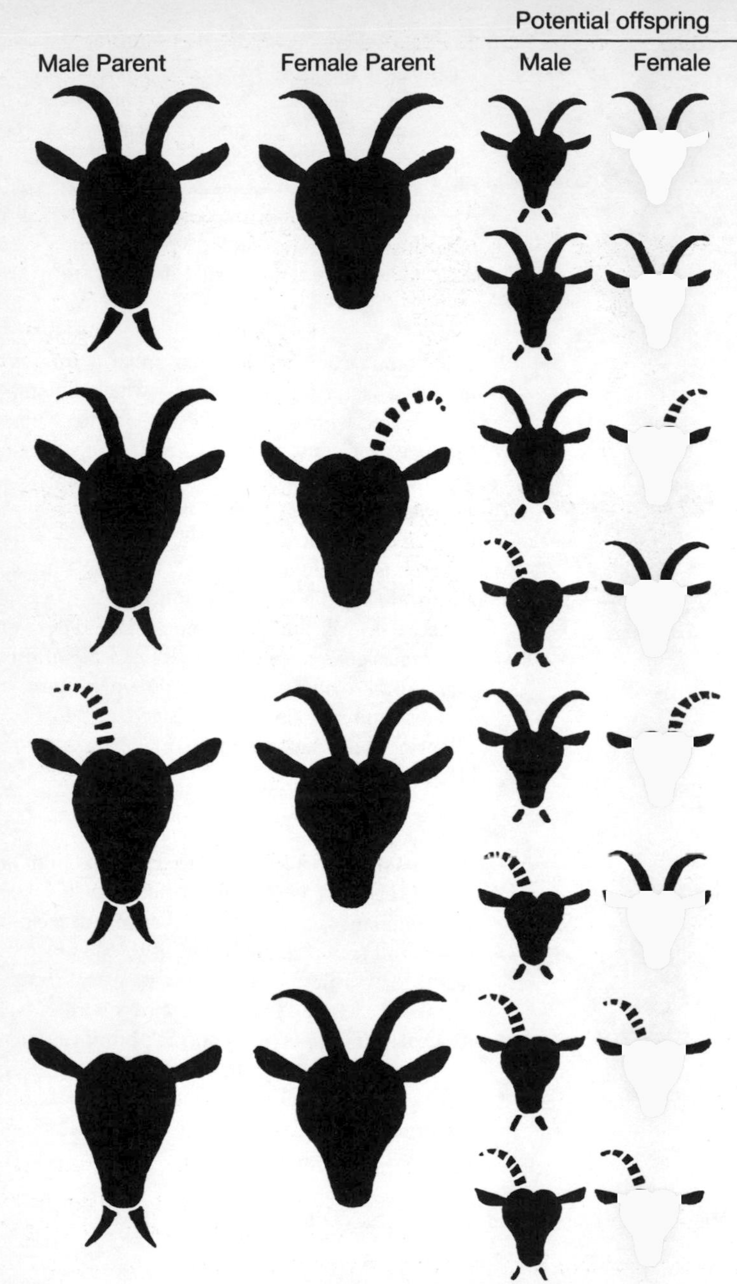

Fig. 16 Hornlessness and the intersex condition (b) Safe matings.

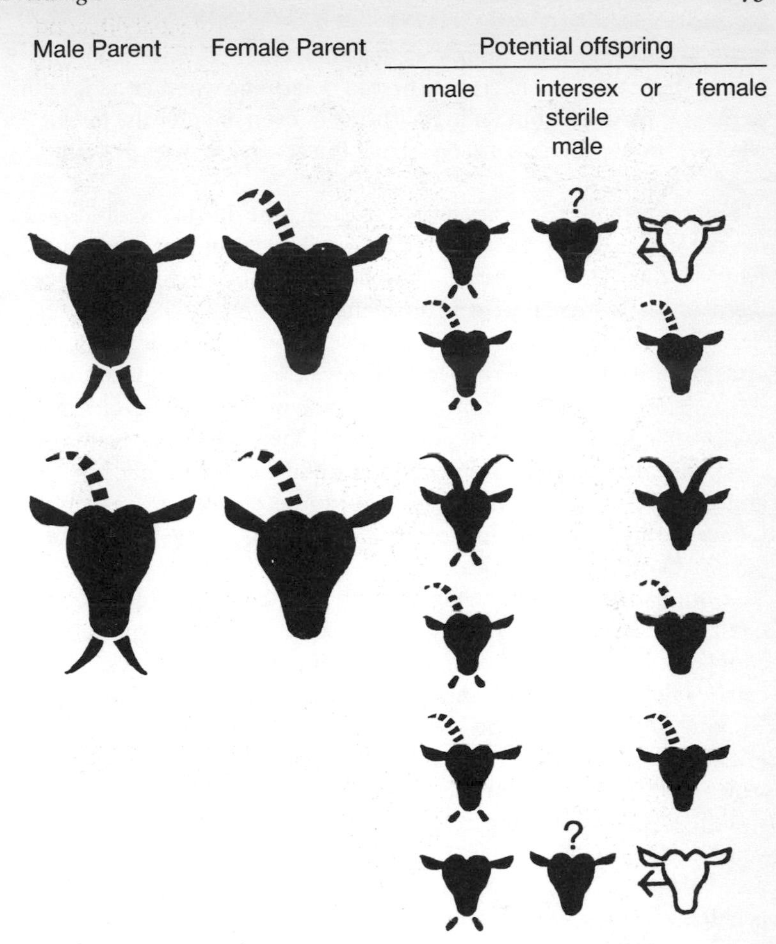

Fig. 16 Hornlessness and the intersex condition (c) Risky matings.

Male infertility

There are many possible causes. If there is more than one male in the herd, and both/all of them are infertile, then there may be management problems: over-use of males; out of season breeding being attempted; a disease problem and/or wrong feeding; or if hand mating is being practised, perhaps the females are not really on heat – though male goats are world-class experts at heat detection! If one male only is having problems, it is necessary first to determine whether he is failing to mount, mounting but failing to thrust or even to enter the female, or whether proper services are occurring but several females are failing to hold to service.

Failure to mount or to attempt to serve may be due to low libido. This may be due to out of season breeding – if light treatment is used on the females, then treat the males as well. Low libido may also be due to disease, the presence of a male higher up the 'pecking-order' or a hereditary tendency. Males with any sort of foot trouble, weak pasterns, arthritis or back pain will understandably be loath to mount a female. If the male is mounting but not entering the female, and he is old enough to protrude his penis (about four months) then veterinary examination will be needed to seek out infection or trauma of the prepuce or penis, and any condition found may be able to be corrected. Entering but failing to thrust may be due to stress, tiredness, strange surroundings, back pain or being over-fat.

If normal services are being achieved but a number of females are coming on heat again, poor sperm quality must be suspected. The simplest cause is over-use, from which it may take six weeks or so to recover fully. Alternatively the male may be too old, too young, or it may be only the very start of the breeding season. Stress may be implicated, but poor nutrition is also a likely reason. It is important to feed working males well, though this may be difficult once their minds are on other things! Besides providing an adequate intake of energy and protein, vitamins, especially Vitamin A, and minerals are important. An incorrect balance of calcium and phosphorus is known to cause breeding problems in both sexes; copper, manganese, cobalt and zinc deficiencies have all been said to reduce sperm formation. Illness, particularly involving fever, may cause permanent or temporary damage to fertility; the overheating of the scrotum in Angora males with a very heavy fleece may impair sperm function for a couple of months after shearing. Veterinary examination of the testes and associated structures may discover some abnormality. The effects of the homozygous polled condition on fertility are stated above.

Female infertility

Once again, it is helpful to determine whether this is a herd or an individual problem. Many of the factors mentioned above – nutrition, general health – are as important for females as for males. There are in addition many problems which potentially cause female infertility. Herd problems are mainly those of management, but if artificial insemination is being used, perhaps poor technique is the cause.

Individual problems can be categorized: if the goat, though apparently on heat, will not stand for mating, then she may not be sufficiently fully on heat, or she may be scared. If neither of these seems to apply, veterinary examination of the vagina and cervix may reveal the cause. Beyond this, it is necessary to record whether the animal is not coming on heat at all, whether she is doing so irregularly, or whether she is in oestrus every three weeks as she should be, but is failing to conceive. Reasons for not coming on heat at all (anoestrus) may include: wrong time of year, already pregnant, heats very hard to see, malnutrition, recent kidding, a heavy milk yield, the results of abdominal surgery, a false pregnancy (cloudburst), genetic conditions such as intersex resulting from the homozygous polled condition mentioned earlier or the freemartin condition, resulting in small ovaries and an inability to breed, which is an occasional result, caused during pregnancy, of having a male twin. The ovaries may be genetically normal, but failing to function. It is essential to discover if the goat is already pregnant, before resorting to hormonal methods of bringing her in season, which would induce abortion in the pregnant animal (e.g. prostaglandin treatment). Pregnancy diagnosis in the goat has improved greatly in recent years. Commercially available milk tests for oestrone sulphate, produced by the placenta, give good results from fifty days after mating; ultrasound scanning, converted into a visual image, can also give excellent results, and is becoming more generally available. A false pregnancy is due to a persistent corpus luteum and can be treated by the vet but can recur and may be hereditary.

Irregular heats have different causes depending on whether they are too frequent or too far apart. It is quite usual to have two heats about a week to ten days apart at the beginning of the breeding season – the second one being fertile. Kids may have short cycles quite normally. Hormonal treatments given in connection with embryo transfer may cause frequent heats. However, there are more sinister causes requiring treatment, such as ovarian cysts, a uterine inflammation or the remains of a mummified kid. If the goat is mated then has another heat more than three weeks after the date, it may be that the embryos have died – changing the male is a wise precaution. It may be that the goat was not pregnant and the next heat was missed. A 'billy rag' is useful, i.e. a rag rubbed on the male's scent glands and kept in a closed jar until offered to a female which may possibly be on heat – the smell will cause tail-wagging if the goat is in oestrus. This is only necessary where no male is kept on the premises. Alternatively the corpus luteum may persist for too long, delaying the cycle. This can be treated by the vet.

An understanding of the hormonal control of the female cycle helps to diagnose any problems which may occur. The use of hormone-impregnated vaginal sponges, coupled usually with hormone injections, which is widely practised to synchronize breeding or to breed outside the normal time of year, may also be used to produce a successful outcome in a difficult breeder. The sponge is impregnated with progesterone, which halts the cycle. Nine days after insertion, Pregnant Mare Serum Gonadotrophin (PMSG) is injected as a source of FSH and LH, together with prostaglandin, to destroy any persistent corpora lutea. On day 12 the sponge is removed, on day 13 the goat will come on heat, and is mated or inseminated that day or on day 14.

Kidding

Perhaps the greatest source of anxiety concerning the goat in kid is uncertainty of the day on which the kid will be born. The average period of gestation, from date of service to date of kidding, is 150 days. The normal range is from 143 to 157 days, which gives the herdsman who has noted service dates plenty of room for anxiety; if service dates have not been noted, the goat will keep her attendant guessing for a month, and spring a surprise in the end. The herdsman who is never taken by surprise is the one who passes his hand inquisitively over the goat's right side every morning and evening, as a matter of routine.

In the pregnant goat, as she approaches her term, the kid can be seen and felt to move in the bulge on her right side. So long as that is the case, there is little likelihood of kidding within twelve hours.

The initiation of birth was described earlier. The kid or kids which have kicked about in a flabby bag suddenly find themselves in a strait-jacket. If they do not lie still, their movements are so little felt by an outsider that they seem to do so. If the goats have been handled regularly there is no possible room for doubt about this change.

Any time from eight to twenty-four hours after the uterus has tensed, in a normal birth, one of the kids is forced up into the neck of it. This movement causes the bulge in the right side to subside, and tilts up the sloping plates of the rump to a more horizontal angle. The change in appearance can usually be noticed if one is looking for it; but if the kids are many or big, the subsidence in the right side is not very obvious. Once this stage has been reached, the first kid may be born within two or three hours.

Other signs of imminent kidding are a widening of the space between the pin-bones; the appearance of deep hollows on either side of the tail, and rapid filling of the udder and teats. But in old goats, and in goats carrying a heavy cargo of kids, the space between the pin-bones may ring a false alarm weeks before kidding; in heavy milkers, the udder is often quite tight and shiny ten days before kidding, and pre-milking may be necessary.

The secretion of oestrogen which tenses the walls of the uterus also affects the behaviour of the goat. A first kidder will often behave very much as if she were in season; an older goat is more likely to appear merely fussy and restless in the shed. But let the goats out of the shed together, and the goat which is about to kid is no longer a member of the herd, no longer follows her leader, but maintains a purposeful and divergent course quite foreign to her normal behaviour. The kidding goat in a free-range herd always gives fair warning, if there is an eye to see it.

Like sheep, goats seem able to postpone or accelerate the birth of their kids to take advantage of good weather; older goats use their discretion in the matter more notably than first kidders.

The first good kidding day which comes within a fortnight in which the kids are due is the normal choice of an experienced mother. A good kidding day is a mild and humid one, with a minimum of wind; lambs and kids are best born in a Scotch mist, where loss of heat by evaporation of the birth fluids is at a minimum.

If the day is good and the terrain at all suitable, the best kidding pen is the secluded hollow that the goat will choose for herself, which is more hygienic than anything under a roof, and freer of hazards than anything with a flat floor and four walls. Most goats will choose a place under the cover of trees or rock; an indiscreet first kidder may fail to do so, and should be encouraged to think again; shelter from crows, showers and gulls is desirable. Otherwise human assistance is best limited to watching from a distance. Company of any kind, even of the dearest, is not desired, and it may be several hours before it is needed out of doors, even if things go wrong.

It is not everyone who has free range for goats, or would welcome the idea of supervising kiddings outside. In fact, a goat may refuse to go out with the herd if she is ready to kid that morning. She should have a clean box, with a good covering of short straw which will not get entangled with her restless movements. Long straw is awkward; and a layer of sawdust under short straw, while it absorbs mosture, needs to be completely covered.

When kidding has started, there is a discharge, at first colourless. When birth is fairly imminent, this changes, becomes thick and white, and is a sure indication kidding has actually begun. The goat paws her bedding; lies down with a sigh or grunt, and perhaps strains very slightly; gets up; walks about restlessly; lies down again and strains. She may look round and 'talk' to the kid, which she can probably at this stage smell – this 'mother talk' is something only heard during kidding and for a few hours afterwards; and once heard, it can never be mistaken. These are all normal signs, and you may perhaps go away and, returning half an hour later, hear the sound which tells you that the first kid is now being born. Some goats are extremely vocal during the final stages, when the serious business of hard straining develops; and the shrieks can be unnerving when heard for the first time. First kidders generally make a lot of noise; older goats probably only give a few anguished grunts at the moment of birth. The goat usually lies down, and it is as well to see that her back end is not pressed against the side of the box. She will probably lie stretched out flat, and push with hind legs against the floor or wall to assist the birth. Some goats drop their kids standing up.

Now, if all goes well, the water bag should appear and burst to reveal a foreleg, followed by another with the top of the nose resting on them (see Fig. 17 for this, normal, and other, abnormal, presentations).

The goat may rest for a moment, then, finally, with two huge heaves, expel the rest of the kid. If there is a hold-up, grasp the kid's legs and, as the goat strains, pull them downwards towards her hocks. The first kidder may not realize that she has given birth or recognize the wet, slimy object as her child. Clean the mucus from the kid's nose and mouth and gently slide it round under the mother's nose; she will immediately lick and talk to it. A second, third or even fourth kid may follow at varying intervals. Presently the afterbirths, one for each kid, a double one for identical twins, will come away; this may take some time, and no attempt should be made to pull or interfere with the process. Keep count: the goat may perhaps eat part of them when you have gone for a cup of tea and to fix her an oatmeal drink; but each placenta has a double set of cords, and from this evidence you know that everything has come away. A retained afterbirth causes trouble later.

Wash the goat's hind parts with warm water and dry them; remove some of the stained bedding, and top up with fresh. Offer her a warm oatmeal drink made by dissolving a handful of oatmeal in cold water to paste consistency, stirring in a kettleful of boiling water, and allowing it to cool a little. Offer it pretty warm, and she will almost surely drink. Some goats like a tablespoon of molasses added.

Meanwhile, examine the kids for abnormalities: a supernumerary (extra) teat, or a double one which appears just unusually thick but has two holes, can be hereditary, and such kids should be destroyed. One of the more obvious forms of intersex is seen on the vulva of an apparently female kid as a small, round, pea-sized object. There are other bisexual forms, with both male and female organs on the same kid. Obviously the only course is destruction, which can be done later; the goat settles more happily if at least one kid, preferably two, stays with her for three or four days.

Make sure the kids have sucked; if the second kid is some time coming, you can draw off some milk into a warmed bottle and let the kid take its first feed from a rubber teat. In this way, there will be no hesitation if and when it is to be bottle-fed at four days. Finally, offer the goat a bran mash, give her a rack full of the best available hay, and leave the family in peace for two or three hours.

The above refers to a normal kidding. If after twenty, or at most thirty, minutes of hard straining, no progress is made, the first kid is probably lying in one of the awkward positions shown in Fig. 17. There is no point in letting the goat exhaust herself and endanger the kids in a futile endeavour to give birth. The novice learns a lot and gains confidence if a vet, experienced goatkeeper or a shepherd delivers these first kids. Next time, given a little common sense and courage, the job will not appear so formidable.

If you are obliged to tackle the situation yourself, get someone to hold the goat if necessary, or you may have to tie her up. Wash your right arm thoroughly to above the elbow, scrubbing nails clean (long nails are impossible; and anyone liable to be delivering kids should have short

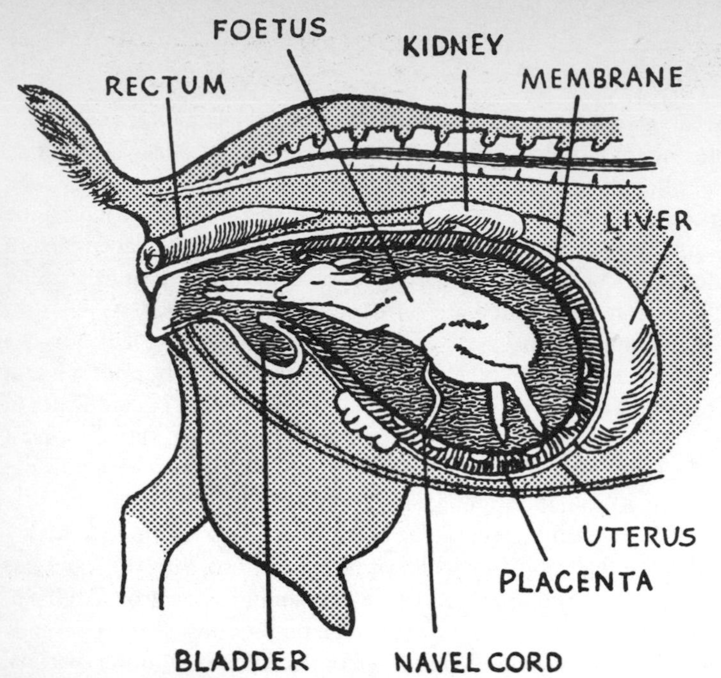

Fig. 17 Parturition diagrams
(a) Normal presentation.

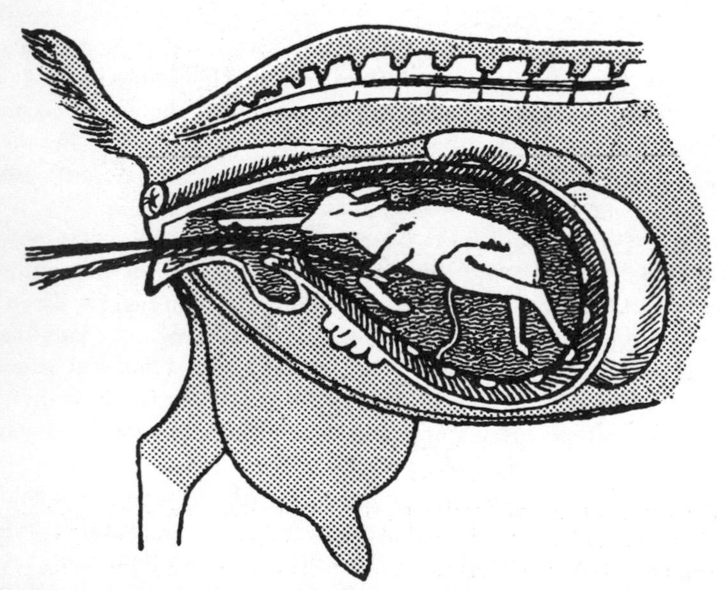

(b) Bring that leg forward – with your finger if possible; the rope shows direction of pull.

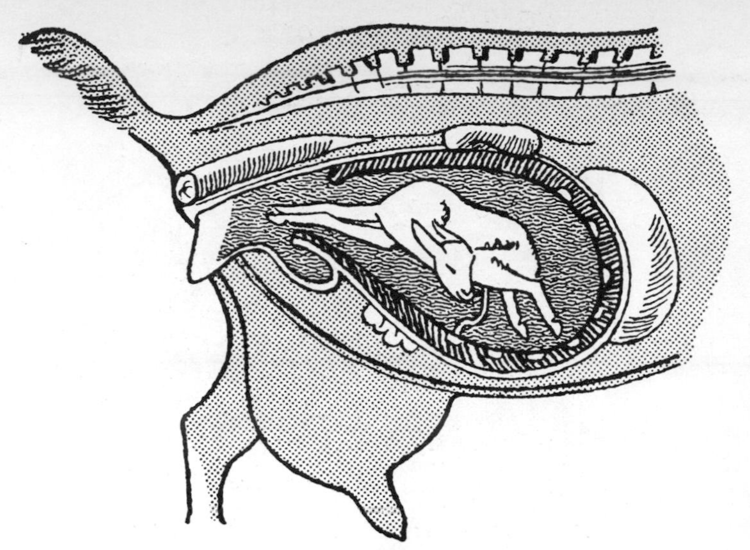

(c) Push the kid right back into the uterus, and bring the head forward on to the front legs.

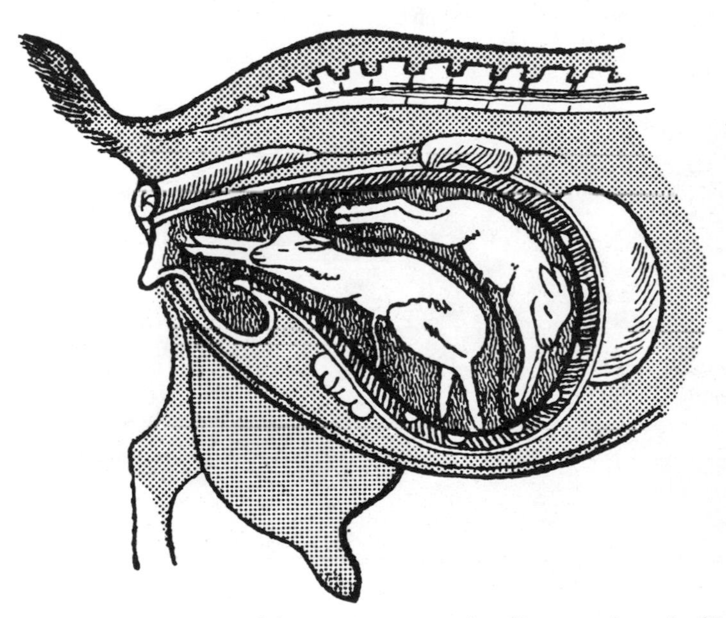

(d) All correct and normal for a twin presentation. But sometimes the hind legs of the second kid get in front of the head of the first. Then you find four legs and a head in the passage. Push it all back, and bring out in the position shown.

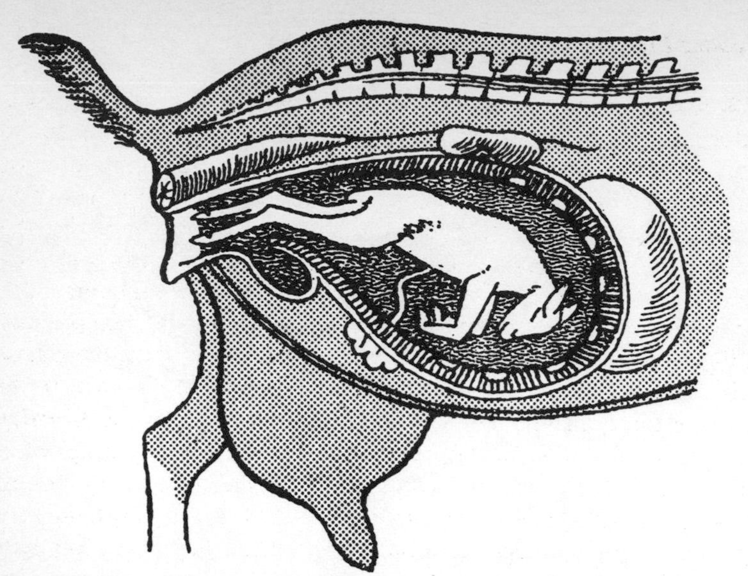

(e) It will come out the way it is; a normal presentation for the second pair of twins; but, especially with an elderly goat, ensure that it doesn't linger too long as the shoulders come through.

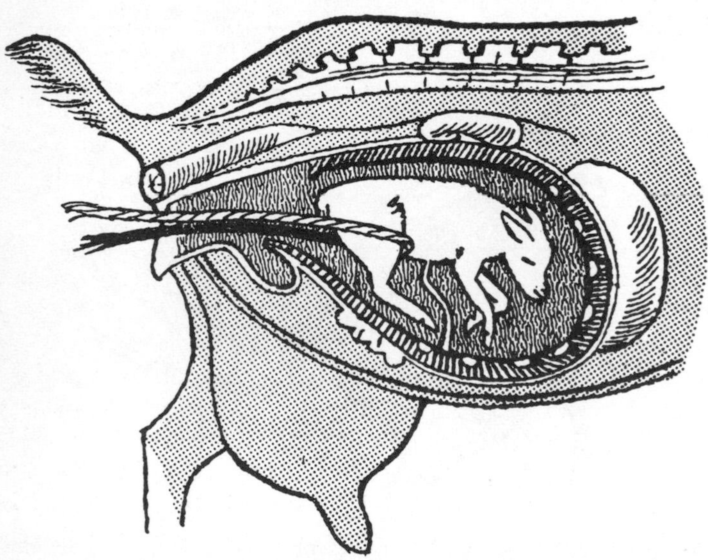

(f) This one doesn't reach the passage at all, but sticks in the mouth of the uterus; push him well back into the uterus and deliver him in position (e). As before, the rope shows what fingers should do.

nails, not cut at the last minute, but smoothly filed down in readiness). After washing, cover your hand and arm in obstetric cream, bunch the fingers together to a cone and, without touching anything on the way, slide gently into the goat. Now feel around, with a finger at first. The goat strains vigorously and you must work in conjunction with the strains. If the trouble is a turned-back foot you may, with a slightly upward movement, be able to bring it forward (Fig. 17b). For an entire leg or head (Fig. 17c) turned back, or other wrong presentation (such as in Fig 17f), you need room to work; and this means pushing the kid, or kids, right back into the uterus. Talk reassuringly to the goat, try to visualize the difficulty, and sort it out intelligently. It is much easier once you get the idea, and if you are able to keep calm and visualize what you can feel. Never pull on anything unless you are absolutely certain you can identify it. Sometimes two kids are in the passage with the fore legs of one and hind legs of another the first thing you find (Fig. 17d); don't pull, push it all back. It helps to handle a young kid, or even a dog, beforehand, so that you can recognize fore from hind, and the position you may have to re-adjust, by touch alone.

Sometimes a goat is all ready to kid, but does no more than stand about looking miserable for some hours. She may have a slimy pinkish discharge or none at all; and she never reaches the hard-straining stage. She eats no hay, and is obviously in trouble. This could be a case for Caesarean section if she has not opened up properly; call the vet.

Disbudding

Examine the kid for horns within twenty-four hours of birth. The horned kid has a slightly flatter head and two curls covering the pinpoint horn buds, which can be felt and, if the hair is closely clipped, seen as small white pimples. The hornless (polled) kid has a slightly domed head and the curls are absent. If horned, make an appointment with a vet in order that they may be disbudded at four days old. Male kids have to be done at two days, but would normally be destroyed or destined for the table or deep freeze at a comparatively early age. All too often the job is deferred for a week or more, which is hard on the kids and makes success less certain. Clip the hair extensively and closely round the horn buds on the third day and, if the kids are to be bottle-fed, remove them from the goat that day and get them used to the rubber or plastic teat. Preparing the head previously and letting the kid go straight to the bottle after anaesthetic for disbudding, reduces stress to the absolute minimum.

Many vets advise a general anaesthetic for kids for disbudding because some kids in the past have died from shock caused by a local anaesthetic. The general anaesthetic also ensures that there is no possibility of a struggle, allowing the kid's head to move. If the disbudding is done not later than the fourth day, the kid will soon recover.

For those interested, the disbudding is generally carried out with an electric iron; no pressure is exerted. After initial cautery, a completely flattened area with a small bit of horn sticking up round the perimeter is left. These bits are then smoothed down and the process is repeated on the other side. After recovering from the anaesthetic, the kid can be given a bottle and be put to rest quietly. The head is very sensitive for the first few weeks: if it gets a knock, the kid will cry out in pain for a minute or two. Subsequently, the head becomes hard and tough as though naturally hornless.

Some breeders used to disbud kids, but this is now illegal. The Veterinary Surgeon's Act 1966, as amended by the Orders of 1988, requires that 'disbudding/dehorning of a goat at any age is to be performed by a veterinary surgeon'. *It is an offence for a lay person to offer such a service or to perform the procedure on his/her own stock.* It is not a job to be postponed. Horned domestic goats, however gentle, are a potential hazard to children, to each other and to those who handle them: the tip of a horn can all too easily blind an eye.

The Ministry of Agriculture, Fisheries and Food published in 1989 (reprinted in 1990) a *Code of Recommendations for the Welfare of Goats*, available free from MAFF Publications (details on p. 324), which confirms that disbudding should be carried out as soon as possible after birth. Reference 4 of the *Code* states that 'only a veterinary surgeon or veterinary practitioner may dehorn or disbud a goat, except the trimming of the insensitive tip of an ingrowing horn which, if left untreated, could cause pain or distress.'

Fleece-bearing goats are frequently left with their horns intact.

Deodorization

A technique has been developed by the Robert Ford Laboratory, Pascagoula, Mississippi, USA, to remove a goat's musk glands, making male goats inoffensive and eliminating the 'goaty' flavour which can affect milk as the result of musk-gland activity in she-goats.

The male goat has a well-known habit of labelling his friends, females and furnishings with his personal stink, by rubbing his head on them. The musk glands are situated in a ⅜- to ½-in (9 to 13mm) wide band immediately behind, and along the inside edge of, the base of each horn. In a naturally hornless animal, it is situated similarly around the bony boss. In a disbudded animal, the gland is to be found where it would be if the animal were horned (see Fig. 18).

Musk glands are present in both sexes. Being activated by the presence of male hormone in the blood, their activity is seasonal in the male and unusual in the female. In an adult, when the normal cover of hair is removed, the glands are seen as an area of thickened and glistening skin; if active, the skin is raised and folded, forming three corrugations on each side; in a kid, the gland is shiny and darker than the neighbouring skin.

The deodorization procedure is a simple extension of the disbudding technique, the glandular area being scorched to a bonelike appearance with a red-hot disbudding iron, by a vet, in the same way as the horn bud. An adult goat, if hornless, can be deodorized by a vet by cauterizing the gland in like manner; if horned, the horns would have to be removed along with the glandular skin, by a long and bloody operation which would be hard to justify. Dehorning of an adult goat is a stressful procedure and should be avoided. In Britain, all such

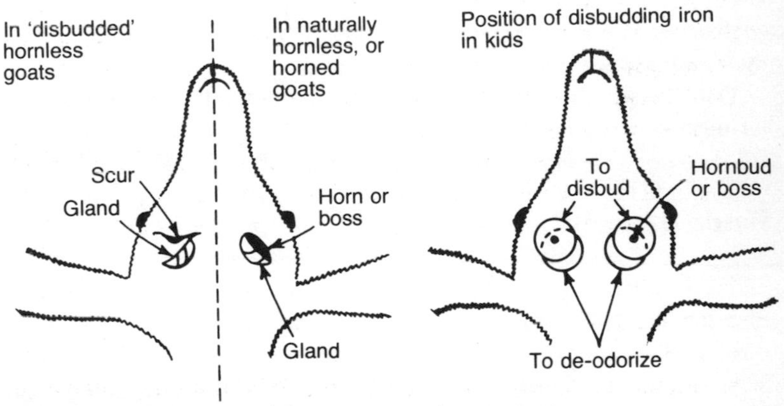

Fig. 18 Deodorization of goats: location of musk glands.

operations must be carried out under general anaesthetic by a vet. In commercial dairy herds and many domestic herds, there is much advantage if all available kids are deodorized when horned kids are disbudded.

There are no serious snags in the procedure. Male fertility is not affected, and kids 'mother' normally. A few animals have musk glands on other parts of their bodies; these can be located by nose, after shampooing the goat, and cauterized in the same way by a vet. A goat suspected of producing goaty milk because of an activated musk gland can be tested by rubbing the glandular area with your hand for a few minutes and sniffing. However, goaty milk is usually the result of the breakdown of fat to form fatty acids, hastened by lipase, an enzyme in the milk.

Destruction of kids

Most male kids and any weaklings, in addition to the abnormal or surplus, will be destroyed at birth or by the fourth day, when the goat has settled down and must be milked dry. Under no circumstances should they be drowned; it is a prolonged and terrible death to which no animal, however young, should be subjected. The preferred methods are use of humane killers or lethal injections. Your vet will advise about this.

Some goatkeepers used to employ chloroform on a pad of cotton wool to put the kid to sleep until it stopped breathing. Chloroform is, however, a very dangerous substance and can be lethal to the goatkeeper as well as to the kid. It is now very difficult to obtain and its use is not advised.

Many people consider it a waste of resources not to rear surplus kids for meat.

Bottle feeding

Kids are either bottle-fed from the fourth day or left for the dam to rear. She must be milked out morning and evening, or you will not have milk through the winter when the kids are weaned. As goats are bred for long lactation, we should make full use of this ability to milk steadily, with kiddings in alternate years. If the kids are to be bottle-fed, get the goat out on the fourth day with the herd; remove the kids out of sight and sound; clean out the kidding pen; and scatter disinfectant around, especially where the kids lie against the door or walls. Plenty of fresh bedding, first-class hay and, if possible, a rack with greenstuff will distract her on return, and she will not fret for the kids. By taking the goat from the kids, never taking the kids from the goat, you have a contented animal settling down to the new lactation; in reverse, you can have a nervous wreck for several days. If the kids cannot be removed to another building, put them in a pen next to the goat so that she can see but not feed them. In a couple of weeks they can run out together without fear that the kids will suck.

CHAPTER TEN

Disease and Accident

Why should the writer of a book on animal husbandry be expected to devote a section to disease? The rest of the book is presumably intended to describe the ways and means of keeping the animal in a healthy and productive state. What more is there to say but to repeat what has already been said? If disease arises in spite of the combined efforts of the reader and writer – as it will – then call in the veterinary surgeon or herbalist, who are technically equipped to deal with the symptoms of disease without removing the conditions that caused it. Their technical equipment is specialized and to be respected; to crib a few of their more popular prescriptions is neither serviceable nor respectful.

If such were the accepted view of the respective roles of stockman and veterinary surgeon, it would be sufficient to supply a few notes on first aid in case of accident and leave the matter at that.

But such is not the generally accepted view. Far from it. The stockman is visualized as defending his charges from a malicious, ever-lurking legion of bacteria, virus and parasite intent on their destruction. The modern stockman is therefore armed with drugs, vaccines, serums and disinfectants to neutralize, suppress and destroy every possible agent of disease. Modern man has made disease after his own image, and is determined to kill it.

This is madness, with a monstrous vested interest behind it, and not a scrap of science. The first fact the scientific student learns about bacteria is that they are neutral – they have no desire nor ability to do anything whatsoever, but that which their environment commands. They multiply only if and when their environment is suitable for their multiplication; they remain dormant or die when it ceases to be so. There is no fight, no argument. A virus is as malicious as a lump of salt. Some forms of virus even look like one: the tobacco virus, for instance, can be crystallized, labelled and kept in a glass bottle for as long as you like. Dissolve it, inject a droplet into the right kind of tobacco plant in the right place, and it multiplies, producing the symptoms of disease.

The conditions under which any given strain of bacteria or virus can multiply rapidly are very highly specialized. In the case of disease bacteria, these conditions only exist in one particular type of animal, or in a few allied species. If the symptoms of disease are very severe, or fatal, the conditions which favour the multiplication of the bacteria concerned must be highly abnormal in the animal species affected, or both the species and the disease bacteria would long ago have perished. On the other hand, disease bacteria which produce only mild symptoms may be able to infect under relatively normal conditions.

In fact, any large animal is a focal centre for a vast and diverse population of micro-organisms and 'parasites' whose way of life is intimately bound with that of their host, and which depend on the survival of their host for their own existence. Some of this multitude of micro- and macro-'guests' are essential to the well-being of their host; others appear to be of little account, one way or the other; some can be helpful or dangerous according to circumstances; others are harmful to the individual they infect, but may act as a useful control of overcrowding when their host is running wild under natural conditions. It seems likely that disease bacteria exist among all these groups.

The relationship between the animal and its 'guests' is well regulated to ensure the survival of both. The more dangerous guests can only gain entry and multiply in the animal body under most unusual circumstances; the less dangerous but potentially harmful guests, by their activity in the animal body, either destroy their own means of sustenance or produce antibodies which control their own multiplication within limits which are safe for the host. The useful members of the community are encouraged to multiply by the circumstances in which they can be useful.

This regulating mechanism works well under normal conditions; if it is to work well for our goats, we must feed, house and manage them to accord as nearly as possible with the life they would lead under favourable natural conditions. In so far as we are obliged to depart from the natural regime, we must expect some breakdown in the mechanism and be prepared to offer additional protection. But the mechanism is generally beneficial: to treat all potentially dangerous bacteria and parasites as enemies is both senseless and expensive.

Legal control of medicines for goats

Our membership of the EC has generated massive alterations in the control of veterinary medicines in the UK, in order that these may be traded throughout the Community. Partly to cope with all this, the Veterinary Medicines Directorate was formed in 1989, the first object being 'to safeguard human and animal health'. To this end, one of their jobs is to license animal medicines. The tests to which manufacturers have to put new drugs in order to obtain a licence is very costly in both time and money. Consequently, there are very few products specifically licensed for use in goats. This is important because drugs not licensed for use in goats can only be given under the direction of a vet, and a standard withdrawal time (i.e. the time after treatment ends during which meat or milk may not be used for human consumption) has to be observed. It is seven days for milk and twenty-eight for meat. The use of drugs in goats is particularly crucial because they are food-producing animals; it became an offence in January 1992 to slaughter, sell or supply for meat any animal containing residues of certain substances. Some products, such as certain hormones, may not be given to meat animals at all, other substances may only be given under the direction of a vet; meat samples will be taken and tested for residues and action taken as necessary by enforcement authorities. Records have, by law, to be kept of all veterinary medicines purchased and given to goats. The headings under which these records are to be made are laid down as follows:

- Name and address of person keeping record
- Date of purchase of veterinary medicine
- Name of veterinary medicine and quantity purchased
- Supplier of medicine
- Identity of animal/group treated
- Number treated
- Date treatment finished
- Date when withdrawal period ended
- Total quantity of veterinary medicine used
- Name of person who administered veterinary medicine.

Veterinary medicine records like this must be kept for all goats, not only for those intended for meat.

Another record which has to be kept in connection with animal health is the Movement Record of all cloven-hoofed animals moving on and off your premises – this will be inspected from time to time by the local enforcement officer. Record-keeping is a major occupation for today's citizen of Europe, and this record is to be extended to cover births and deaths under the Registration with MAFF of all goatkeeping premises, and marking of stock prior to movement (see Chapter 17). Movements of all stock by motor vehicle must also be recorded.

In the field of veterinary medicines, the VM Directorate carry out post-licensing surveillance of drugs by welcoming any reports from users under the Suspected Adverse Reaction Surveillance Scheme.

It is unlawful for anyone to sell veterinary medicines without a licence to do so and the VMD oversee the control of agricultural merchants by the Royal Pharmaceutical Society. Drugs are categorized as Prescription Only Medicines (POM) which can be obtained only from a vet, and Pharmaceutical Merchants' List (PML) which can be bought from licensed farm supply shops, etc. It is greatly to be hoped that the latter will not be terminated under EC law, so that all medicines will have to be obtained from a vet.

Medicated foodstuffs are strictly controlled: this control is also overseen by the VMD. Veterinary homoeopathic medicines have not escaped the EC legislators; a full product licence will be needed to sell these to treat food animals, though they will still be able to be prescribed by vets, just as for conventional medicines. The withdrawal times for homoeopathic veterinary medicines will be waived when the principle is present at a concentration of one part per million or less (potency 4C and upwards).

Animal welfare is a considerable factor in promoting animal health, and it is a major concern of the UK government to try to upgrade animal welfare standards in the rest of Europe to a level equivalent to our own. Welfare legislation covering animals in transit, at market and at slaughter will be dealt with in Chapter 13.

The EC is, of course, concerned with free trade; unfortunately it looks as though we are going to trade freely in animal diseases as well as animals. Legislation has been put in place to try to ensure the good health of goats being traded across Europe, and being brought into Europe from Third World countries; this will be covered in Chapter 17. British goatkeepers will need to be more vigilant.

Look out for these signs

	Reassuring	*Worrying*
Head	Bright-looking and alert	Dull or distressed
	Nose and mouth clean	Dribbling or nasal discharge
	Eyes clean	Eyes discharging, inflamed

	Ears in normal position	Ears drooping which are normally pricked
Body	Coat with a healthy sheen	Coat dull, hair falling out
	Coat smooth	Signs of skin irritation
	Skin supple	Skin tight, 'hidebound'
	Suitably covered with flesh	Too fat or too thin
	Faeces normal pellets	Faeces lumpy, runny, mucus-covered
	Urine normal, passed easily	Urine abnormal, passed with difficulty
	Behaviour normal	Behaviour abnormal
	Appetite good	Off food
	Female: no vaginal discharge	Vaginal discharge, other than normal
	udder even and soft	Udder hard, lumpy, milk abnormal
	Male: genitalia normal	Testes uneven, swollen, signs of pizzle rot
Temperature	101·5–103°F	Widely outside this range
Respiration	18–25/minute	(may be more in hot weather)
Pulse rate	70–90/minute	(may be more in hot weather)
Legs and feet	No lameness	Lameness
	Clean legs	Swollen/hot joints
	Well-shaped feet and hooves	Misshapen feet, signs of rot
General	Everything normal	Any signs of abnormality.

Some diseases and conditions of goats

Abortion

There are many causes of abortion, and only some of them are infective agents. These, however, may be dangerous to pregnant women, who should not attend kidding or aborted goats. It is also inadvisable for pregnant women to drink unpasteurized goats' milk, even though any health risk may be extremely small.

Acetonaemia

Signs: goat goes off eating concentrates; may continue to eat roughage. Milk falls, irregular cudding. Droppings become dark, covered in sticky mucus. No fever. Positive urine or milk test.
Occurrence: usually within 2–3 weeks of kidding. Goat is usually a heavy milker.
Prevention: correct feeding. Ensure enough energy, but not overfat goat.
Cure: cobaltized salt for 10 days; increase roughage; injection of glucocorticosteroid will stimulate appetite, ½lb (227g) flaked maize night and morning will probably help. Encourage eating.

Anaemia

Not a primary disease, but a sign of cobalt deficiency (pine), or of worm and fluke infestation or other conditions.

Blindness (contagious ophthalmia)

Signs: one or both eyes watery; cornea becomes cloudy; white spot develops at cornea. Spot enlarges, becomes reddened and may rupture, causing loss of the eyeball.
Occurrence: germ probably spread by flies, but also carried from infected eyes to healthy eyes by the tips of flapping ears while goats are feeding at trough. Usually more than one case in a herd.
Treatment: ophthalmic ointment or antibiotic powder in eye are commonly used. Best treatment would seem to be subconjunctival injection of an antibiotic – long-acting tetracycline possibly best.

Bloat (blown, hoven, tympany)

Signs: tightly inflated flanks; misery; collapse.
Occurrence: on lush clover and grass (especially wet grass) pastures in spring and autumn, gorging on anything unsuitable and after raiding the food bin.
Prevention: feed hay and allow an hour's cudding before turning out. Re-seed pasture with 'Herbal Lea' Mixture (see p. 274).
Cure: drench with vegetable or other oil 6–8fl oz (150–200ml) for an adult goat, 2fl oz (50ml) upwards for kids. Walk the goat about, massage flanks. After this treatment wind is usually expelled from either

or both ends and goat rapidly returns to normal. If not call vet as an emergency.

Brucellosis

Infection of sheep and goats with *B. melitensis*, the cause of Malta Fever in humans, has not been recorded in the UK for many, many years, but it is important in Southern Europe, where the EC are trying to eradicate it. The UK has to prove to the EC, by blood-testing a certain number of goats and sheep each year, that it is free of *B. melitensis*. The disease became officially notifiable on 1 January 1993.

Caprine arthritis encephalitis (CAE)

A disease of major importance in many goatkeeping countries including France. The virus was discovered in the UK in 1982, and is closely related to the maedi-visna lentivirus in sheep. Goats infected remain so all their lives, and are infectious to others, though the majority exhibit few if any signs, and can be detected only by blood test. The disease is slow to develop and infected animals may not give a positive blood reaction for many months, or even some years.
Signs: clinical signs are said to appear in some goats when 25 per cent of the herd becomes infected. There are four forms of the disease. Arthritis: not before six months, and usually in adult goats. The membranes of joints and tendons become inflamed and produce excess fluid. The knee (carpal) joints are commonly affected – the disease is known as 'big-knee' in several countries. There is no fever and appetite remains good, but wasting and lameness gradually take over. Hard udder: a firm, swollen udder in freshly-kidded goats; little milk produced, and what there is is not let down, though it is normal in appearance. There is no response to treatment and though the udder may soften, tissue changes in the udder mean that milk production stays low. Uneven development of the two sides of the udder has been seen frequently in French cases. Pneumonia: progressive pneumonia with a chronic cough and weight loss. Encephalitis: nervous signs, including paralysis of hind limbs, can occur in young kids, but has not been reported in the UK as yet.
Occurrence: blood-testing has shown a low level of infection, between 4 and 8 per cent in the UK, due no doubt to the fact that most owners have had positive reactors destroyed to control the spread. This is in

sharp contrast to situations which occurred in the USA, Australia and France, for instance, where the majority of herds became infected, necessitating drastic eradication measures. The virus lives in one type of white blood cell and is transmitted by body fluids containing them, notably colostrum and milk, though saliva, blood, etc., are also minor agents of spread. CAE has been called 'show-goat disease' due to the unfortunate infection of many kids at shows, by being fed on milk which had been in the milking competition weighing bucket, before the situation was realized. Feeding pooled colostrum in herds at home was thought to have been the major factor in the dramatic spread across the USA. Kids from infected mothers must be 'snatched' at birth before they can be licked or fed. Direct contact between goats can transmit the disease, but probably not rapidly; however precautions are obviously sensible.
Prevention: most organizers of goat shows, and owners of male goats at public stud, insist on a sight of a negative blood-test certificate dated during the previous twelve months, though even this obvious measure is objected to by a few 'head-in-the-sand' goatkeepers. Because it is possible for infected goats to give an occasional negative test result (sero-conversion), however, it is much better to have a means of certifying that the whole herd is regularly tested and negative. To this end, the British Goat Society runs a 'Monitored Herd Scheme', membership of which involves annual testing of all goats over twelve months and record-keeping of the herd, with disposal of any positive goats, though goats are free to attend shows, etc., as the owners wish. The Ministry of Agriculture's Sheep and Goat Health Scheme, on the other hand, which offers CAE-free Accreditation, combines testing with strict measures to avoid contact with non-accredited stock. It is interesting to note that Accredited herds show almost no positive tests, while infection in other tested goats continues to run at the same level.
Cure: none.

Caseous lymphadenitis (CLA)

Signs: abscesses of the lymph glands, caused by a bacterial species, infection with which results in thick cheesy pus being formed. Bacteria from this pus remain viable in the environment for months, and enter new hosts through small cuts, etc. The introduction of one infected animal can result in a high number of abscessed animals in the herd two or three years later. Usually the superficial lymph nodes of the neck and

chest are affected but internal abscesses can also occur; some as big as grapefruit have been recorded.
Occurrence: once absent in the UK but rife on the Continent, this disease was unfortunately introduced from Germany and confirmed in this country in 1990, and has since been imported again from France. The disease has an incubation period of around three months and between abscesses goats can appear normal. Blood tests are available through the Ministry of Agriculture.
Prevention: extreme care in buying-in goats. If possible insist on blood tests. Watch for abscesses and, if in doubt, ask for veterinary confirmation.
Cure: surgical treatment of abscesses may be indicated.
Note: very rarely, this infection has been passed to humans.

Coccidia

See Internal Parasites.

Colic

Signs: spasmodic pain in the digestive tract; goat half rises, sighs and groans, goes down again, suddenly stops eating and looks anxiously about; may be tympany of the left flank; increasing distress; leads to death if unchecked.
Occurrence: commonest in youngstock; conditions favouring worm infestation; excessive feeding of concentrates, access to cold water immediately after concentrate feed; in kids, after swallowing bottle teat; after poisoning.
Prevention: correct feeding and constant access to water; feed the flock roughages together; hold the teat when bottle feeding.
Cure: ½ pint (0·3 litres) of vegetable oil (for adults, less for kids), followed by 1 glass spirits in 2 glasses of water, repeated hourly until pain subsides.

Cuts and Wounds

All cuts but the very smallest are best stitched, as this reduces the healing time. Most cuts are contaminated to some extent, and this both delays healing and provokes the goat to attack the wound, further delaying healing. Antibiotics reduce the infection which accompanies contamination, and speed healing. Tetanus antiserum is also indicated, unless the goat is vaccinated. A local or general anaesthetic may be needed for the cleansing and suturing of large wounds.

Cuts on the udder or teat exposing the teat must be stitched. The act of milking after suturing may re-open the wound, and the use of hollow teat cannulae with removable plugs to let the milk drain should be considered. Intramammary penicillin injection (both sides) is indicated to prevent the development of mastitis.

Dermatitis

See Eczema.

Diarrhoea ('Scouring')

Not a disease, but the sign of disease caused by worms; coccidia; enterotoxaemia; salmonella and other bacteria; poisoning; indigestion; over-feeding.

In kids, it may be due to irregular feeding; or to the use of unsterilized feeding-bottles.

Eczema

Several forms of skin disease occur in goats, two of which have the appearance of eczema.

Contagious pustular dermatitis (Orf)
Signs: pimples about the nose and mouth, less often about the eyes, anus and hoofs, turning to watery blisters; then to sticky and encrusted scabs. Also on udder of suckling goat.
Occurrence: a local infection of sheep country, commoner in eastern districts. Young animals most frequently affected; but all ages liable on first contact with infection.
Prevention: home-bred stock in an infected district have natural immunity. Vaccine will give protection to others.
Cure: difficult. Dress with antibiotic sprays.
Note: zoonotic (i.e. infection can be passed to humans).

Goat pox (True virus goat pox does not occur in UK)
Signs: pimples turning to watery blisters; then to sticky and encrusted

scabs on the udder. If scabs removed before 'ripe', weeping sores remain.
Occurrence: common, with a wide variation of severity, depending on the virulence of the germ and the resistance of the stock.
Prevention: scrupulous dairy hygiene and isolation of affected milkers, which should be milked last and the hands disinfected thereafter, will help to control the spread of severe outbreaks.
Cure: time and gentle milking. A cream with antibiotic and cortisone may help.

Enterotoxaemia

This is the worst goat killer, though true incidence is unknown.
Signs: they vary somewhat according to the strain of bacteria concerned; but are usually consistent in any one flock. Staggering gait and loss of motor control is one of the less common signs; sometimes the goat is blown; there is always extreme misery; and, almost always, peculiarly evil-smelling diarrhoea. Coma and death within 24 hours are the normal sequel.
Occurrence: ineradicable bacteria, indigenous to goats, and to all other domesticated grazing stock and their pastures, produce the poisons responsible, when conditions in the digestive tract deprive them of oxygen. In goats, a big feed of lush, wet grass, or of concentrates and water, or a real belly-stretcher of milk, all produce the airless pudding in which the bacteria start poison production. Goats quickly build up resistance to the poisons produced in small regular quantities; it is the sudden change of weather and diet in spring and autumn, and the accidental gorges of goats in mischief, that cause most trouble.
Prevention: correct feeding on a bulky, fibrous diet. Biennial treatment with vaccine gives protection.
Cure: on first acquaintance, enterotoxaemia is usually fatal before diagnosed. Once the characteristic symptoms have been seen, they are more easily recognized when they recur.

Immediate recourse to the hypodermic syringe will usually save the goat. Two injections, each of 10,000 units of penicillin, four hours apart, or a single injection, of 300,000 units of slow-release veterinary penicillin, is likely to be successful; sulphamethazine (dose according to weight) gives good results. Whatever treatment is to be used, it must be kept in permanent readiness for immediate use. Delay proves fatal. Give Clostridium perfringens type D, antitoxin and painkiller.

Ergot

See Poisoning.

Fluke

See Internal Parasites.

Foot-and-mouth disease

Symptoms: sudden lameness; small, dribbling blisters on the tongue, inside of lips and on the palate; also where hair joins hoof and between the claws of the hoof. Can be confused with contagious pustular dermatitis.
Occurrence: rare. Inform the police. Do not move animal or try to treat.

Foot rot

Signs: slight and increasing lameness in one or more feet; hoof uneven, with stinking matter oozing between the outer horn and inner soft structures.
Occurrence: among goats on sodden pastures or floors; but the disease is either a regular visitor, or an infected sheep or goat introduces it. The bacteria concerned can remain latent for long periods in the hoof but cannot survive a week on the ground.
Prevention: dry floors and the avoidance of sodden pastures; regular exercise on hard ground. Regular trimming of feet, vaccination with foot-rot vaccine.
Cure: pare the foot level; remove loose horn and, very gently, as much dead matter as possible. Soak all four feet for 20 seconds in a 'Dettol' solution (4 tablespoons to 1 pint [0·6 litres] water). Repeat daily till sound. Then, once every 10 days for a month, dip the feet briefly in a copper-sulphate solution or proprietary foot-rot wash. Keep the infected feet, if not the goat, out of contact with the communal pasture, using doll's Wellington boots, perhaps. A 10 per cent solution of formaldehyde is quite effective as a foot bath. Even a small amount in a tin can, in which the feet can be immersed above the hair line one at a time, is sufficient. Goat should then be stood on concrete or wooden floor until solution is dry. Repeat weekly for 3–4 weeks.

Fractures

A fracture (if skin not broken) below elbow of fore leg or below hock of hind leg can usually be treated by plaster of Paris, depending on value of goat. Surgery may be successful in other cases. Consult vet.

Gas gangrene

Occasionally infects a deep penetrating wound, producing a foul-smelling bubbly discharge. Death may follow in a few days. At an early stage, antibiotics and serum may control the infection. Vaccination against clostridia generally may prevent this disease.

Goat pox

See Eczema.

Hoven

See Bloat.

Internal parasites of goats

(Details of this section supplied by Dr E. M. Abbott, Technical Development Manager, Hoechst Animal Health.)

Gastro-intestinal roundworms
Signs: main signs are diarrhoea and wasting; milk production will also be depressed. If *Haemonchus* infection (barber's pole worm), then the main signs will be bottle jaw and anaemia.
Occurrence: acute disease seen in late summer and early autumn especially in young animals. Older animals usually have more chronic disease.
Prevention: by treating all stock early in the grazing season and repeating this treatment at intervals of 4–6 weeks depending on stocking rate and availability of pasture.
Cure: alternatively, treat and move to 'clean' grazing e.g. hay or silage aftermath.

Treat on housing at the end of the grazing season.

Females should be treated around the time of parturition and during

early lactation as they are immuno-suppressed during this period.
Note: Many of the gastro-intestinal roundworms of goats also infect sheep.

A very worrying situation is developing world-wide, in that gastro-intestinal roundworms of sheep and goats are developing resistance to all the known worm-medicines. The problem tends to be worse in goats, which do not develop immunity to worms as sheep do. In the UK resistance has been to 'white drenches' like Panacur, but in 1992 resistance to Ivermectin was found on a goat farm in Scotland. Recommendations for avoiding resistance are: change to a wormer with a different active ingredient (p. 203) once a year but not more often; do not under-dose by underestimating goat's weight. Panacur, give 1·5–2 times the sheep dose rate; Nilverm Gold, give 1·5 times the sheep dose rate, not more as it is toxic; Ivermectin, use at sheep dose rate. Follow maker's instructions. Do not drench unnecessarily – ask your vet for a strategy; control worms by management not dosing if possible: i.e. moving to clean pasture in July and grazing land with cows and horses to destroy goat worm larvae. Treat bought-in goats with two different wormers and keep them off grass for twenty-four hours, after second dose. Suspected resistance should be checked by lab. If all else fails, keep goats housed.

Lungworms
(a) *Large lungworms* *Signs*: cause chronic cough and failure to thrive.
Occurrence: disease usually confined to young animals.

Infection accumulates in the animal during the summer.
Treatment: with anthelmintic. The prophylactic scheme suggested for gastro-intestinal roundworm control will prevent build-up of lungworm infection also.
(b) *Small lungworms* *Signs*: rarely associated with disease in goats in the UK.
Occurrence: an intermediate host (snail) is involved in the life cycle. Development of small lungworms in the goat is slow and as immunity is also slow to develop; adult goats are more likely to have heavier infections.
Treatment: with anthelmintic (not all the licensed goat wormers are effective against some small lungworm species; consult vet).

Tapeworms
Signs: tapeworm infection is usually symptomless and experimental

studies have failed to demonstrate clinical effects even with heavy infections.
Occurrence: the intermediate host is a forage mite; goats become infected when they ingest the mite. Infection is more common in the first year of life and young goats will pass tapeworm segments in their faeces during the summer months. Infections are short-lived and persist for only a few months.
Treatment: with an anthelmintic which has activity against tapeworms.

Liver fluke
Occurrence: although in theory goats could become infected with the liver fluke, *Fasciola hepatica*, there are no recent references to infection with this parasite in goats. Liver fluke infection is common in sheep on the west side of the United Kingdom in areas where land is poorly drained and of poor quality. This habitat encourages the survival of the intermediate host, a snail. Liver fluke infection in sheep causes wasting, and severe anaemia with bottle jaw occurs in heavy infections.
Treatment: there is an excellent, safe flukicide available for sheep (Fasinex, Ciba-Geigy) but as it is not licensed for use in goats, a veterinary surgeon's advice should be sought.

Coccidia
These are uni-cellular parasites which infect the intestine of young animals.
Occurrence: disease occurs mainly in kids and the main signs are severe diarrhoea which may have blood in it. There is rapid weight loss, dehydration and death if left untreated. Disease outbreaks are more likely to occur under conditions of poor husbandry when kids are kept in unhygienic conditions.
Prevention: depends on ensuring that pens are kept clean and dry. Feed and water containers should be raised to prevent contamination by infected faeces.
Treatment: specific anticoccidial drugs such as sulphamethazine are available from veterinary surgeons.
Note: Anthelmintics (wormers) are *not* effective against coccidia.

Johne's disease

Signs: loss of condition; occasional scouring, becoming more frequent with bubbles of gas in the droppings; appetite usually good; emaciation;

Table 15 Wormers licensed for use in goats*

Trade Name	*Active Ingredient*	*Withdrawal Time*	Spectrum of Activity *Gastro-intestinal roundworms*	*Ovicidal*	*Lungworms large*	*Lungworms small*	*Tapeworms*	*Use*
Panacur (Hoechst)	Fenbendazole	Flesh 14 days Milk 72 hours	√	√	√	√		Safe at all stages of pregnancy
Nilverm Gold (Coopers Pitman-Moore)	Levamisole	Flesh 3 days Milk 24 hours	√	×	√	×	×	Safe at all stages of pregnancy
Oramec Drench (MSD AgVet)	Ivermectin	Flesh 14 days Not for use in goats producing milk for human consumption or within 28 days prior to start of lactation	√	×	√ (adult only)	×	×	

√ = active
× = activity

*Data based on ABPI Data Sheet Compendium, 1989–90

weakness; death; may take several months to run its course.
Occurrence: infection may be picked up from cattle or sheep.
Prevention: management to prevent spread of infection.
Cure: none.

Lactation tetany

Signs: anxiety; uncontrolled movement; staggering; collapse; convulsions – all in rapid succession – death if not immediately checked.
Occurrence: commonly when put out on to pastures in the flush of spring; also in autumn towards height of breeding season.
Prevention: correct feeding and management; especially, provision of magnesium at times of likely deficiency.
Cure: immediately inject about 100ml of 25 per cent magnesium sulphate solution subcutaneously in loose skin behind elbow. A surgeon *might* inject a weaker magnesium solution intravenously.

Lice

Signs: skin irritation; rubbing, scratching etc.; bald patches; lice seen on skin.
Occurrence: very contagious; common in goats in poor condition, especially those with little access to open air, and to rainfall.
Prevention: regular treatment.
Cure: there is no louse powder licensed for use in goats, or any other domestic animals. If an unlicensed product is prescribed for use on milking animals, milk must be withheld from human consumption for seven days. Pour-on louse preparations; Parasol (Ciba-Geigy) and Ovipor (C-Vet) are licensed for goats and there is no withholding time for milk if the instructions are followed correctly.
Note: louse infestation in Angora goats is likely to result in the animals' being barred from entering show-grounds. Pour-on preparations are much more suitable than powders for a goat with a heavy fleece.

Listeriosis

This is a serious disease of many animals, goats and humans among them, due to a bacterium of widespread occurrence in soil, etc., particularly associated with eating contaminated silage. Infection can have various results – encephalitis, septicaemia, abortion, and so on.

Interest has been focused on it due to public health considerations. Milk can carry the infection, which would multiply in soft mould-ripened cheeses which were low in acid – but there is evidence that it is just as likely for contamination to get on to the cheese during manufacture. One case of encephalitis in a woman who had contracted listeriosis from soft goats' cheese contributed to 'listeria hysteria' in 1990 and did the goats'-milk industry much harm.

Louping ill

Signs: dullness and fever; followed by tremor, muscular spasm and bewildered gait; often by collapse, paralysis and death.
Occurrence: on tick-infested pastures in Scotland, North of England and North Wales during the period of tick activity – April and August–September. Home-bred youngstock are only slightly affected, seldom developing the nervous symptoms; home-bred adults are very seldom affected at all; newcomers of all ages are liable to serious infection during the first tick season.
Prevention: home-bred stock develop natural immunity. In-bought stock from a tick-free area should be inoculated with 5ml louping-ill vaccine in July and March of their first year.
Cure: shelter, and quiet and careful nursing, will often save the patient. The presence of a congenial companion in the sick bay is a help.
Note: The virus excreted in goats' milk is a danger to human health. Goats in tick areas should be vaccinated.

Mange mites

Sarcoptic mange

This is caused by *Sarcoptes scabiei*, and is relatively common, occurring in all age-groups. The mites, which are difficult to see in skin scrapings as they may be very few in number, burrow into the skin causing terrible itching. The skin becomes raised, red and hairless round the eyes, ears and nose. The damaged skin may suffer bacterial infection. Infestation can be passed to other goats by direct or indirect contact. Eventually the whole body surface may be affected; the goat may lose condition and the milk yield fall. Veterinary treatment is required.

Psoroptic mange

This is caused by *Psoroptes cuniculi*, which infests the ears, inside, and may cause head-shaking and scratching. It is not very common, but infested mothers may pass it on to their kids. If the mites cause the ear-drum to rupture, ear disease and paralysis of the face may occur. Veterinary treatment should be sought.

Chorioptic mange
This is caused by *Chorioptes caprae*, which infests mainly the lower parts of the legs, causing scabs but not much itching. Bacterial infection of the damaged skin can occur and requires antibiotic treatment. The mites can be killed with a suitable veterinary shampoo.

Demodectic mange
This is caused by *Demodex caprae*, which live and multiply in the sebaceous glands of the skin, causing nodules in the skin, from which can be expressed cheesy material containing the mites. Itching is not produced. Goatlings are the most likely to exhibit nodules, following infection as kids. British Alpines are said to be particularly prone.

Mastitis

The goat is liable to all forms of mastitis that affect the cow; and to black garget of sheep. Several different diseases with different causes are grouped under this head. Only in a minority of cases is mastitis in goats associated with specific bacteria; many cases are not associated with bacteria at all; some forms are highly contagious to all goats; others are contagious only to goats in an unhealthy condition; others again arc not infectious at all.

Acute mastitis
Signs: misery; udder hot, hard and very tender; milk clotty and often blood-streaked; appetite lost; pupils of eyes narrowed to slits. In the worst cases, the udder may putrefy and slough away, and the goat die in high fever.
Occurrence: in goats fed a concentrate diet, especially during the first fortnight after kidding; and following injuries to the udder. Highly contagious to goats similarly fed.
Prevention: correct feeding and management; let kids suck for first four days of lactation.
Cure: antibiotics, supportive therapy.

Sub-acute mastitis
Signs: as for acute mastitis; but appetite near normal, and less distress.
Occurrence: as for acute mastitis; but at any time during lactation.
Prevention: as for acute mastitis.
Cure: as for acute mastitis.

Chronic mastitis
Signs: lumpy udder; occasional clots in the milk; occasional off-flavours; no notable discomfort or distress.
Occurrence: as for acute mastitis.
Prevention: as for acute mastitis.
Cure: as for acute mastitis.

Summer mastitis
Signs: dejected or anxious expression; slitty eyes; lumpy udder, which may develop into acute inflamed udder, which putrefies or shrinks and hardens into a dead lump. Often fatal.
Occurrence: mainly in goatlings or older kids during summer; occasionally in dry goats; only affecting flocks in which large quantities of concentrates or concentrated succulents are fed to youngstock.
Prevention: correct feeding and management.
Cure: as for acute mastitis.

Black garget
Signs: as for acute mastitis; but udder rapidly becomes black and gangrenous, either sloughing away or hardening into a dead lump. Often fatal.
Occurrence: occasional in all goats after a wound on the udder, particularly a penetrating wound; in goats suckling a lamb.
Prevention: remove barbed wire from fences; control dogs; bottle-feed orphan lambs.
Cure: a job for the vet, who may be able to get the antiserum in time; and otherwise may succeed with antibiotics, or iodized oil suffusion. While awaiting the vet's arrival, carry out hot and cold fomentation, massage and frequent stripping.

Sub-clinical mastitis
Signs: udder and milk appear normal, though pathogens are present. There may be a reduction in the yield of milk and in the milk solids content, and the milk will not keep so long. Possibly 4 to 6 per cent of

udder-halves are infected in this way.
Cure: antibiotic treatment.
Notes (1): Milk from infected udders will have a high bacterial count and will probably fail the sample tests which are to become law on 1 January 1994 when milk is to be sold for human consumption (see Chapter 11).
(2) There are at present no intramammary antibiotic preparations licensed for use in goats, which are available in the UK. This means that the EC standard withdrawal time of at least seven days after treatment ends must be imposed. However, it is known that the times taken for antibiotics to clear from goats' milk differ from those for cows' milk – sometimes more, sometimes less. This is a matter for concern in view of EC milk hygiene rules requiring that milk must in future be sampled for residues, and in any case the health of consumers, and the usefulness of the milk for processing, are both put at risk by contaminated milk.

Metritis

Inflammation of the womb.
Signs: slowly developing feverishness to point of acute misery and collapse, usually with increasingly foul-smelling discharge from the vulva.
Occurrence: not before kidding, unless the kids are dead. Within a few days of kidding, especially if kidding is assisted, or cleansing retained, or incomplete kidding (retained kid).
Treatment: antibiotics.

Milk Fever

Signs: inco-ordination; staggering; collapse and inability to rise.
Occurrence: maybe a month before kidding to a week after kidding; but usually within two days of kidding.
Prevention: correct feeding will not prevent this disease.
Treatment: 200ml of 20 per cent calcium borogluconate at blood heat subcutaneously. Inflation of the udder is unnecessary.

Mycoplasma agalactia (contagious agalactia)

Not thought to occur in the UK, but officially notifiable from 1 January 1993. Imported goats must be tested for it within five days of arrival.

Ophthalmia

See Blindness.

Orf

See Eczema.

Pink Milk

A trace of blood in the milk is common shortly after kidding, especially in first kidders, and is due to the bursting of small blood vessels in the udder. If it is not accompanied by other symptoms, this is innocent. If the condition persists it indicates lack of calcium (or an excess of phosphates, i.e. concentrates) in the diet.

Pneumonia

Signs: signs of fever, i.e. depression; off food; standing with back arched; coat on end; breathing faster than normal, may be laboured and with grunt.
Occurrence: after stress.
Cure: antibiotics – tetracycline, etc.

Poisoning

Generally speaking, goats are fairly immune to semi-poisonous plants if on free range with plenty of variety; but there comes a time when forage is scarce, or a gate left open, so they gorge on something unsuitable with dire results. Prompt action is absolutely essential: decide what the trouble is and tackle it immediately. *Poisonous Plants*, published by HMSO, is useful for identification.

The principle is the same in all cases. First remove the goat from the poison, then the poison from the goat. Having removed the goat from the poisonous shrub, food bin or whatever, house her in a warm place covered with a rug, and, if extremities are cold, massage and surround with covered hot-water bottles. Otherwise, keep her moving to stop her chewing the cud. Unless the goat is vomiting, carefully give her lots of strong tea (not hot!) as the tannic acid will render many plant poisons insoluble. Also tea, and coffee, are useful stimulants. However, do not give tea in cases of acorn poisoning as tannic acid is the poison in acorns and you would be adding to the trouble. Giving the goat a mixture of egg, sugar and milk may soothe her stomach. As always, be careful not to pour fluids into the lungs; stop dosing at once if the goat coughs, then recommence when she stops coughing. Consult the vet if signs of poisoning persist.

A liquid paraffin drench is frequently advisable. Dosage varies from 6fl oz to ½ pint (170–284ml) or more, depending on severity of poisoning and size of goat; kids need 0·2–0·4fl oz (5–10ml). Small soft-drink bottles are convenient for administering this. Do not dose if goat is vomiting.

After-treatment consists of bran and molasses mashes; no concentrates at all; hay; fresh water offered or changed frequently.

Special indications for particular vegetable and mineral poisons, and for overeating of concentrates in a raid on the food store, are as follows.

Acorns (unripe)
Treatment: ½ pint (0·3 litres) liquid paraffin drench; laxative food. Do not give tea in this case, but coffee.

Arsenic
Signs: slobbering; scouring; thirst; attempts to vomit; onset rapid.
Occurrence: found in weedkillers, sheep dips, etc.
Treatment: 6oz (170g) Epsom salts in ½ pint (0·3 litres) water as purge; follow with liquid paraffin drench and white of egg in milk.

Azalea
Treatment: as Rhododendron.

Beet Leaves
Treatment: suspension of ground chalk in milk as antidote, or calcium borogluconate injection to combat hypocalcaemia if necessary.

Concentrates
Prevention: it is a golden rule that bins or bags containing concentrates should never be accessible to stock. A goat is exceedingly artful, and will even manage to undo many types of catch before pushing up the lid of a bin. There should never be the least possibility of this happening.
Treatment: speed is vital. Purge with 6oz (170g) Epsom salts in ½ pint (0·3 litres) water; get this down somehow. Withhold water, which swells

grain and makes things considerably worse. If the poison is grain, the vet will give an intravenous injection of Parentrovite (Vitamin-B complex). Treat for shock (rugging, warm stabling); follow later with smallish liquid paraffin drench. If ruminal impaction is acute, treatment for acidosis, loss of rumen bacteria and dehydration can be attempted by the vet.

Conifers (except Yew, q.v.)
Treatment: Epsom-salts purge, size of dose according to severity of poisoning.

Ergot
Signs: lowered condition; cold and loss of sensation in the extremities.
Occurrence: among goats confined to pastures on which ergot, a fungus growth on flowering grasses and cereals, is prevalent.
Prevention: keep goats away from such pastures.
Treatment: none is effective.

Hemlock (roots in winter)
Treatment: strong, sweet, black coffee or tea with 1fl oz (28ml) spirits as stimulant; keep goat moving.

Herbicides and Pesticides
Signs: typically, sprays cause violent black scour and attempts to vomit.
Occurrence: many goats have been killed when browsing near the hedge in a field next to one being sprayed.
Treatment: 6oz (170g) Epsom salts in ½ pint (0·3 litres) water as purge; later, smallish drench. After such poisoning, it is vital to keep the goat off any form of greenstuff for several weeks, or even months, until all after-effects have disappeared. (See also Arsenic.)

Laburnum
Treatment: as Hemlock.

Laurel
Treatment: as Rhododendron.

Lead
Signs: colic; intense pain; goat grinds teeth and slobbers; often followed by delirium and blindness.

Occurrence: found in paint, linoleum, roofing felt, etc.
Treatment: there is one specific and highly efficacious antidote, which should be given in cases both where symptoms are apparent, and where poisoning is merely suspected. This is calcium disodium versinate solution; it is given intravenously.

Mangold Leaves
Treatment: as Beet leaves.

Pesticides
See Herbicides and pesticides; also Arsenic.

Rhododendron (and similar shrubs)
Signs: vomiting; spitting out greenstuff; pain.
Treatment: first aid; 6oz (170g) Epsom salts in ½ pint (0·3 litres) water; 1 pint (0·6 litres) liquid paraffin (for demulcent effect); very strong, sweet black coffee.

Or: first induce further vomiting with 2 tablespoons melted lard and 1 tablespoon bicarbonate of soda, shaken together in a warm bottle and carried to the goat house in a jug of warm water to stop it solidifying. Follow 1 hour later with 1 cup molasses and 1 tablespoon bicarbonate; 2 hours after this, strong tea or coffee as above. Obtain veterinary help.

Sheep dip
Often contains arsenic, *q.v.*

Exposure to organophosphate dips can cause abdominal pain, diarrhoea and severe nervous signs, coma and death. The vet may give atropine intravenously.

Snake bite (adder)
Signs: shock; if snake bite suspected, look for swelling and two small punctures. Adder bite is seldom fatal, except on udder.
Occurrence: the adder or viper is the only poisonous snake in Britain; from springtime onwards it likes to lie sunning itself, usually on piles of stones, flat rocks or dry banks. Black-and-silver colouring (brown just before casting skin) with V markings all the way down back; quite unlike beige, unmarked, non-poisonous slow worm. Goats less likely to be bitten than dogs since, on finding a snake, they stand still, and the shy snake slinks away.
Treatment: rub permanganate-of-potash crystals into wound; treat for

shock. (This may also save a dog, if done within 20 minutes or so.) Call vet for anti-venom.

Weedkillers
See Herbicides and pesticides; also Arsenic.

Yew
This dangerous plant can kill within minutes.
Treatment: unlikely to be effective; but try usual poison routine, with extra-large liquid paraffin drench (up to 1 pint [0·6 litres] for a large goat) and 1fl oz (28ml) spirits. Keep goat moving; follow up with molasses and strong tea or coffee as for Rhododendron, but omitting the bicarbonate of soda.

Pregnancy Toxaemia

Signs: dullness, stupidity; loss of appetite; collapse; death – all within 12–48 hours.
Occurrence: occasionally in goats on all standards of nutrition whose diet fails to improve in quality during the last six weeks of pregnancy; lack of exercise increases liability.
Prevention: correct feeding and management, with special reference to steaming-up and regular exercise.
Cure: inducing abortion with cortisone may save goat. Nothing guaranteed, but 2oz (56ml) glycerine mixed with 2oz *boiling* water, given as drench twice a day, works well with ewes.

Ringworm

Signs: round, dry, scabby patches appear on the skin. Contagious to humans; infections can remain in buildings and reinfect.
Prevention and cure: vet will supply the feed additive Fulcin, which should be fed to all goats.

Scrub out sheds and boxes with washing soda in strong solution; treat wood with creosote.

Scrapie

This is a progressive fatal degenerative disease of the central nervous system of sheep and goats. It has some relation to similar diseases in cattle, mink and man. Much research has not yet finally solved the riddle of how it is caused – part an infectious agent, part genetic. The disease has itching and nervous signs. It is not common, but attention has been drawn to it by the EC's insistence that it should be made notifiable from 1 January 1993. Furthermore, since that date, breeding goats may not be traded between EC member states unless at least two years of freedom from scrapie on the vendor's holding can be officially proven (see Chapter 17).

Tetanus

Signs: tense, anxious posture, head held up, neck outstretched, ears pricked, top lip retracted in snarl, tail elevated; slight tympany on left; legs rigid, like rocking horse.
Occurrence: following any wound, whether deep or not, if contaminated by the tetanus germ – even following disbudding a kid. Occasionally spontaneous, with no known injury.
Prevention: either tetanus antiserum at time of injury, or incorporate tetanus in the vaccination programme.
Cure: massive doses of tetanus antiserum to counteract the toxin, and penicillin to kill the germ – recovery doubtful. Vet will give injections to control muscle spasms. Fluids must also be given.

Tetany

See Lactation tetany.

Ticks

See Lice; and Louping ill.

Toxaemia

See Pregnancy toxaemia.

Tympany

See Bloat.

Worms

See Internal parasites.

Foot Trimming to Avoid Infection and Malformation

The dairy goat lacks the exercise on rock which keeps the wild goat's feet in shape, and it becomes essential to cut away the outer horn and very carefully trim the hoof so that it stands squarely on a flat surface. Feet should be attended to every four weeks, starting when a kid is about six weeks but omitted in the last two months of pregnancy. All too often feet are neglected, the outer rim turns over, grit and mud accumulate and lead to infection, which can even be transmitted to the rest of the herd. Misshapen feet are sometimes impossible to reclaim after a long period of neglect, so this regular trimming is a necessary

Plate 13 This pleasant British Alpine milker would stand better for her photograph if her hooves were trimmed by regular exercise on hard ground. 'Dutch clogs' are apparent on her near hind hoof, but the 'back on her heels' stance is typical of the goat whose feet are maintained by the paring knife instead of exercise. She was bred by Mr Egerton and gave 4,200 lb in 365 days.

routine exercise. If started early in life, with a sharp knife or foot-rot shears, and provided that care is taken over the job, which should be done when the hoofs are soft after a period on wet pasture, the goat accepts the operation, instead of fighting against it. There is a right and a wrong way to tackle it, and a demonstration by an experienced goatkeeper is the sensible approach.

Alternative medical treatment

There is no doubt that herbalism, homoeopathy and other alternative treatments have their successes and their place. However, they must be administered by experts, other than for first aid purposes, in the same way as conventional medicine. Homoeopathic veterinary surgeons with an interest in treating goats are available.

Table 16 First-aid kit for the goat house

A sharp knife and scissors	Liquid paraffin
2 in and 1 in (5 and 2·5 cm) Prestoband bandages	Drenching bottle
	Hypodermic syringe
Adhesive plaster	Enamel bowl
Surgical gauze	Clinical thermometer
Cotton wool	Foot-rot shears
Surgical thread and needle	Bicarbonate of soda
Boiled brine (or Dettol)	Veterinary penicillin*
Tincture of iodine	Sulphamethazine*
Acriflavine emulsion	Chloromycetin (aerosol spray)*
Hydrogen peroxide	A goat blanket

*Veterinary surgeons are allowed to supply these items only for use under direct supervision

CHAPTER ELEVEN

Milking Practice and Dairy Produce

Legislative control

Nobody wants to give their own family food poisoning, or to produce dirty milk for animal feeding, but the legislation controlling the production and sale of goats' milk and dairy produce applies only to that being sold or supplied (i.e. given away in the course of business) for human consumption.

In 1988, MAFF produced a code of practice, 'The Hygienic Production of Goats' Milk', based on an earlier similar code prepared in Scotland (Code number BL 5677). This grew out of concern among Environmental Health Officers and others that the increased volume of goats' milk being offered for sale represented a potential health hazard, because it was not covered by the Milk & Dairies Regulations which controlled cows' milk production.

In 1989, a Committee on the Microbiological Safety of Food was set up, known as the Richmond Committee after its chairman, the terms of reference of which was to advise government on matters relating to the microbiological safety of food and on such matters it considered needed investigation. Part I of the committee's report was published in 1990 and part II in 1991. Part II dealt with milk and one of the committee's recommendations to the government was that 'milk and milk products originating from sheep and goats should be subjected as soon as possible to controls similar to those which apply to cows' milk and products made from it'. The government readily agreed to this, and planned to use powers given to it under the Food Safety Act 1990 to put the controls in place. However, there was an unexpected delay in the shape of EC proposals for Community regulations covering the production and placing on the market of milk and milk-based products, 'milk' being that from cows, goats, sheep and buffaloes. It is not possible for one EC member state to bring in new legislation on a subject covered by EC proposals.

Meanwhile, the Food Safety Act 1990 itself, and Regulations made under it, had profound effects on goatkeepers who sold milk or other foods. It came into force on 1 January 1991, and has many new features compared to earlier food law. The idea is that everyone in the food chain should be responsible for their actions, all the way from farmer to shop or restaurant, so the goats, their milk and its processing into cheese, etc., packaging – even the water used in the dairy – all have to be up to scratch, or else! Everything can be officially inspected, and there are severe penalties for committing offences, the main ones being: selling, or possessing for sale, food which does not comply with food safety requirements; rendering food injurious to health; selling, to the purchaser's prejudice, food which is not of the nature, substance or quality demanded; falsely or misleadingly describing or presenting food. The Law is enforced by local authorities – Environmental Health for hygiene matters, and Trading Standards for labelling – it is wise to consult them on whether one's selling activities comply. Officers can enter food premises, inspect food, detain or seize suspect food and get it condemned. They must be given reasonable information and assistance. They can inspect processes and records, and take away food samples for testing. Having done so, if they are not satisfied they can issue an improvement notice, a prohibition order or an emergency prohibition notice, depending on how bad conditions are. If prosecuted your defence is that you 'took all reasonable precautions and exercised all due diligence to avoid committing that offence'. Much has been written on the defence of due diligence; courts will decide each case but it is assumed that a small business might not be required to undertake all the precautions which would be expected of a large one. All this, and the possibility of two years in prison or an unlimited fine, have made some goatkeepers give up selling milk without waiting to see if their premises are satisfactory.

Regulations made under the Food Safety Act also affect goatkeepers with food businesses, however small. Labels have to bear datemarks. In keeping with EC law the old 'sell by' date was terminated on 1 April 1991 and now any produce has to be labelled with a 'best before' date or a 'use by' date. It is up to the producer to decide which, but the 'use by' mark followed by the appropriate date is for highly perishable foods which could become a food safety risk. The date must show the day and month, and optionally the year as well. It is illegal to sell food after its use-by date. Other foods have 'best before' and a date composed of day, month and year or 'End 1993' for instance, for longer-life foods. Frozen foods and long-ripening cheeses also have to be datemarked. Storage instructions such as 'below 5°C' should also be on use-by labels. Labelling generally is a bit of a minefield, what with the need for your name and address, the metric quantity, an ingredients list, whether it is made of 'unpasteurised milk', and the name of the food as well, of course, many of which have to be in the same field of view. If you are making claims such as 'low fat', 'reduced fat', etc., the amounts of fat in a portion or a certain weight are also legally laid down. For pasteurized milk or dairy produce, there will be the approval number of your goat-dairy premises too. Sometimes quantities of some ingredients, and/or nutritional data, are required. Labelling regulations do seem to be subject to change; a no doubt very confused Trading Standards Officer should be able to set one straight.

Another Regulation, an Amendment to the Food Hygiene Regulations, concerns the temperatures at which certain foods are stored and delivered. Many dairy and meat products have to be at a maximum of 8°C, while especially vulnerable foods such as cut ripened soft cheeses and cooked ready-to-eat foods must be no higher than 5°C (except in small local delivery vans). It is not necessary to have a refrigerated van if you can keep the temperature down by using cold-boxes – but remember if you deliver to shops it is part of their 'due diligence' not to accept milk, etc., which is too warm.

Your food premises cannot be inspected if nobody knows where you are, and the Food Premises (Registration) Regulations 1991 require all food premises to be registered with their local Environmental Health Office. Registration is simple, free of charge and cannot be refused. You have to register if you use your premises for a food business on any five or more days in five consecutive weeks. Storage of food is a business, so if you have milk or meat for sale in your freezer, you have to register. If, however, you sell fresh milk on the day it is produced on only four or fewer days in five consecutive weeks you may not have to register. If in doubt ask your local Environmental Health Officer. Once registered, you have to notify any changes which occur in your business.

It is required that anyone who works with food shall choose and obtain suitable training for all personnel, such as a course in basic food hygiene, lasting about six hours in all. Many goatkeepers have successfully done this.

To return to the EC milk hygiene directive – this will be implemented by national legislation to come into force on 1 January 1994. This will set the standards on which premises are judged by visiting inspectors. The requirements are these; for those selling milk and dairy produce into the wholesale market, i.e. to shops for resale, or milk to cheesemakers, etc., and directly to the consumer:

Milk production

Goats must undergo regular health inspections in regard to several specific items, including treatment which might leave residues in milk.

Conditions of goat housing, milking area, and handling, cooling and storing the milk must all be suitably designed, constructed and maintained hygienically. Milk rooms must be easy to keep clean and floors must have suitable drainage; they must be well lit and ventilated, supplied with potable water, well away from lavatories and dung heaps and goat houses, be protected from vermin and contain suitable refrigerators. Movable outdoor milking bails can be used under certain circumstances. There must be isolation facilities for infected goats. Animals of all kinds must be kept away from milk rooms. The hygiene of milking is laid down, also that of equipment and personnel. Milk must usually be cooled to 6°C. Official checks must be made that milk is not being watered down. Milk being processed must be checked for bacterial level and residues. Milk to be drunk unpasteurized has to be of a very high hygienic standard.

Milk processing

This includes the preparation of pasteurized milk for drinking, and dairy products such as cheese, yoghurts etc. From 1 January 1994, processing premises must be approved by their local Environmental Health Office. The Approval Number must be placed in a specific way on all produce. To meet approval, processors must use only milk of the hygienic standards laid down. Premises must be adequate in respect of working areas, walls, floors, ceilings, doors, ventilation, lighting, handwashing facilities, cleaning and storage facilities, refrigeration, handling methods, pest control, waste disposal, water supply etc. Equipment must be suitable for the processing carried out. Pasteurized drinking milk must be automatically packaged and pasteurizers must be of an approved design with safety devices. The business must monitor its own products to ensure hygiene standards are met; the monitoring system must be approved. Staff must be trained and new employees must have medical certificates. The somatic cell count, laid down for cows' milk, is not to be established for goats' milk until 1 January 1998 as current research in the UK, goverment funded, has shown that the

Plate 14 Mrs P. Grisedale's milking parlour, Llanon, Cardiganshire. She was the pioneer of the first goat co-operative and promoter of a milk (spray) dryer scheme. (Photograph: Michael Murray)

level in goats' milk appears to be able to be much higher than in cows' milk, without implying mastitis, and this is being further investigated.

The EC Directive continues for many pages, laying down standards for every aspect of processing, wrapping, labelling, transport and storage of milk and dairy produce. There are microbiological standards for cheese and other products. The precise nature of the health mark, to be placed on each wrapper, bearing your registered approval number, is also laid down. Documents must accompany wholesale deliveries.

In its wisdom, however, the EC has recognized that many goats'-milk enterprises are small affairs, supplying only local consumers, directly, via the farm gate or a personal market stall. Providing enterprises supplying only direct to the consumer are free of brucellosis and tuberculosis, they are to comply with national hygiene standards in each member state, and the EC Milk Hygiene Directive does not cover them. These national rules, in the UK, are likely to be the same for small enterprises as for large processors supplying the wholesale market.

Maintaining good hygiene

This requires facilities, know-how and constant painstaking application of effort. The basics are set out in the legislation and those who sell milk must follow them; those who are producing milk for their loved ones will want to do so too.

Clean bedding in the goat house goes a long way to producing clean goats, but the dry nature of goat droppings tends to mean a plentiful supply of dust containing myriads of coliform bacteria, which will readily contaminate the milk. So milking must not be done in the goat house, and goats which have long hair around their udders must be kept trimmed. This long-hair problem is not confined to Toggenburgs and Golden Guernseys; many high-yielding British milkers with a Toggenburg in the pedigree have inherited it.

Milking rooms must be used for no other purpose. An impervious floor should slope to a trapped drain so that all washing water, etc., can be suitably disposed of (milk must not get into waterways or public drainage systems). The internal surface of the milking room must be impervious and easy to wash – internal woodwork is out nowadays! There must be adequate natural and artificial lighting and ventilation, plenty of potable cold water and separate hand-washing facilities; this includes a clean towel.

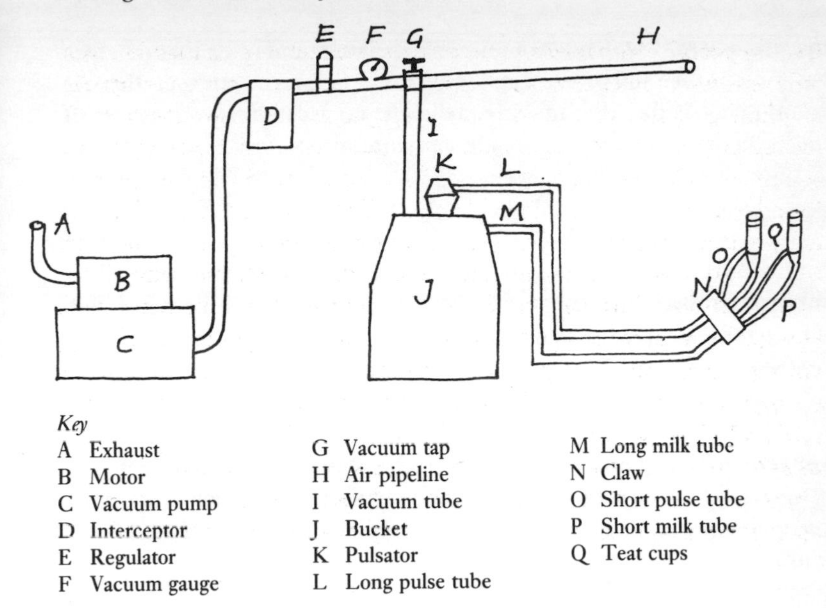

Key

A Exhaust
B Motor
C Vacuum pump
D Interceptor
E Regulator
F Vacuum gauge
G Vacuum tap
H Air pipeline
I Vacuum tube
J Bucket
K Pulsator
L Long pulse tube
M Long milk tubc
N Claw
O Short pulse tube
P Short milk tube
Q Teat cups

Fig. 19 Bucket milking system.

Once the goats are milked the milk is taken to a milk handling room, which once again must be clean and used for no other purpose, adequately lit and ventilated and of impervious construction easy to keep clean – if there are window-ledges they must slope inwards and downwards. The floor should be non-slip, sloping and drained. There must be hand-washing facilities and separate facilities for washing the milking equipment, draining and storing it. The milk must be stored refrigerated, below 4.5°C or frozen, below −18°C. After milking, the milk must be cooled to 4.5°C within one hour (the EC say 6°C) which cannot be achieved with tap water alone in hot weather.

Milking equipment should be of stainless steel, must be cleaned and sterilized with MAFF-approved dairy chemicals after use, and used for nothing but milk.

Milking

Only milk from healthy goats should be retained. Others should be milked last and their milk properly disposed of. Prior to milking, teats and udder can be washed and dried with disposable towels or wiped with proprietary impregnated wipes. If hand-milking into an open pail, discard any milk which becomes contaminated, and put a lid on before carrying to the milk room. If machine milking, make sure everything is running correctly. After milking, a disinfectant teat-dip may be used. If udder cream is needed, apply after milking not before.

Filtration

Immediately after hand-milking milk must be filtered through a disposable filter in a dairy strainer. Milking machines include filters which must be used as instructed.

Cooling

Immediately after filtering, milk must be cooled; small amounts with water followed by refrigeration or freezing, larger amounts in a refrigerated bulk tank. Never put warm milk straight into the refrigerator or freezer.

Packaging

Disposable ready-printed packaging is available; plastic bags with a heat sealer, waxed cartons with a closure strip, and screw-top plastic bottles with a self-adhesive label. All can be used for freezing. For all operations with milk and dairy produce, clean overalls and head covers should be worn. Large enterprises will need automatic packaging, but for small and medium-sized businesses manual filling will be practised. Do not touch the inside of the packaging.

Refrigerating and freezing

Accurate thermometers must be kept inside the equipment to show that they are running cold enough: 4·5°C is ideal for refrigerated milk and −25°C for extended storage of frozen milk.

Testing

EC requirements are for a fortnightly test, but UK requirements for small enterprises have not yet been stated. The code of practice requires freshly packaged unpasteurized milk to have not more than 1,000 bacteria/ml, a very clean standard.

Transportation

This must keep the milk at its storage temperature.

Cleaning and disinfecting equipment

Dirty utensils are the most common cause of poor quality milk. All equipment must be rinsed immediately after use, preferably with lukewarm water. Never let milk dry on. Washing comes next, either with a one-step dairy detergent/sterilizer at the recommended temperature (MAFF approved) or by warm detergent water first, then a second step to sterilize, either with a chemical sterilant or boiling water/steam. Do not, however, mix a detergent with a disinfectant such as hypochlorite, as the detergent will de-activate the sterilant. Finally rinsing with cold water, to which 1fl oz (25ml) of approved sodium hypochlorite solution per 8¾ gallons (40 litres) of water can be added. Never dry milking equipment with a cloth, leave it to drain. If wished, it can be resterilized with boiling water before use. Use recommended sterilants only, and follow the maker's instructions, being careful that the pack has not passed its expiry date. It is very important to leave the chemical in contact with the equipment for long enough. Use a stiff brush for washing.

Personal hygiene

Always be careful to protect milk and produce from contamination. Imagine that bacteria are everywhere, like a thick layer of dust! Always keep the hands and forearms clean by frequent washing. Wear protective clothing – to protect the produce! Skin cuts must be covered with coloured water-proof dressings. Do not smoke, spit or chew! Do not work with milk or dairy produce when suffering from infectious illness.

Hand milking

Milking should be at regular intervals – as near twelve hours as may be – otherwise, the milk drawn after the longer interval will be relatively low in butterfat.

The actual routine of milking will depend upon the standards of

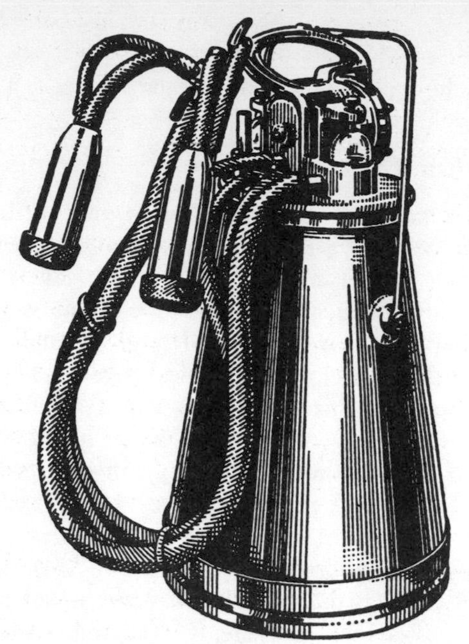

Fig. 20 Gascoigne goat unit.

hygiene required and individual circumstances. But it must be a rigid, orderly routine, involving the minimum of fuss and bother.

The goat requires at least five minutes' warning that she is about to be milked, in order that the hormonal let-down mechanism may release all the milk available when required. During the ensuing five minutes, there must be no distracting novelty in procedure, or the let-down will be slowed. There must be no undue delay once the process is under way, for if the goat is disappointed in the expected moment of her milking, the let-down will have ended.

This mechanism must be remembered when arranging the routine for milking the first goat: the fact of the first goat being milked is adequate warning to the next on the list; but the first goat is entitled to some warning herself. Where udder washing and drying is practised, this procedure will constitute fair warning. Where it is not practised, some routine which will command the goat's attention must be incorporated in the programme.

The goats must be milked in a set order. In a flock which enjoys some measure of social organization, the order should be that of the social hierarchy – an order which a little observation of the flock on range will soon reveal. A flock queen will not give of her best if a couple of goatlings are milked before her. Her protests will be broadcast to the whole flock, who will then be more concerned with her complaints than with the milking routine.

An earnest silence upon the part of those doing the milking is not necessary. Peace in the milking shed there should be; but peace and silence are not synonymous. Men and women have not been singing at the milking for the past thousand years or so merely to exercise their lungs and lose milk. A consistent singer can make the milk drip from the goat's or cow's teats merely by singing the accustomed songs. For a goat distracted by a breach of routine, or by a tender udder, the accustomed song or the regular patter of soothing nonsense is the best aid to let-down available. There is every reason for excluding strangers from the milking shed, but the silent milkmaid is a grim, unhistorical figure with little to recommend her.

The one sound which is quite unforgivable in the milking shed is a sneeze: apart from hygienic considerations, this sound resembles the goat's alarm signal, and invariably (and literally) causes consternation.

The actual process of milking is a knack which is quickly acquired if the basic principles are understood. Anyone who is accustomed to milking cows must realize that they are handicapped by a habit of milking unsuitable for the goat, and pay particular attention to the principles of goat milking.

Grasp the teat lightly in one hand and press the hand gently upwards towards the base of the udder, so filling the teat with milk. Then close the index finger tightly around the neck of the teat, with the hand still pressing gently upwards. This action will trap the milk in the teat. Now close the other fingers in succession tightly around the teat, so forcing the milk down and out. Release the grasp of the teat, and relax the upward pressure. Then repeat the process with the other hand and the other teat.

As the udder empties, the upward pressure of the hand becomes firmer and the initial movement of the milking hand becomes a gentle upward punch. When the flow begins to subside, let go both teats; and, with both hands, gently massage from the top and back of the udder down towards the teat, two or three times. Continue milking until the flow subsides once more; repeat the massaging movement – and so on until no more milk is drawn. The goat should be milked right out to maintain butterfat content of the milk and to sustain the yield – but 'stripping', as practised on the cow, will distort a goat's udder in a very short time.

Many goatkeepers are bad milkers; a novice may be better advised to follow this advice rather than doubtful example.

In a goat with very short teats it may be impossible, especially for a man, to get more than one finger of the hand around the teat. In this case, on the first upward pressure of the milking hand, trap the neck of the teat between the index finger and the knuckle of the thumb; then extract the milk from the teat by rolling the index finger down the teat while keeping the teat pressed against the thumb.

There is a strong temptation, in the circumstances, especially after the first flush of milk has been drawn, to 'stretch a point' and grasp a portion of the udder above the neck of the teat, so as to bring all four fingers to bear. This procedure is dangerous, for it is quite easy to crush the substance of the udder and force a portion of it down into the teat canal, with disastrous consequences.

It is of no importance whether the goat is milked from the right side or the left, provided she is milked consistently from the same side. A goat with an udder of reasonable shape and capacity cannot be milked from behind without great difficulty and damage to the udder. Scrub goats with low yields, and goats with cleft, pendulous bags, can be so treated without harm or trouble. The advantages of this method will be conspicuous to the intelligent beginner. A restive goat cannot kick over her bucket or put her dirty foot in it when milked from this angle. She can, however, drop dung pellets into the bucket, and even make her water into it if the milker is insufficiently alert. The method, though popular with Mediterranean goatkeepers, is not recommended; but really evil-tempered goats are exceedingly difficult to milk in any other way.

The nervous, fidgety, 'ticklish' goat, usually a first kidder, or one suffering from udder injury, can be dealt with by milking with one hand and gripping the goat about the thigh with the other, so as to encompass the hamstring. Compressing the hamstring will smother a kick, even after it has got under way.

More violent objectors may need to have both hind legs, or all four legs, strapped into position on the milking-stand. Even then, it may be necessary to sling a strap under the goat's belly to stop her lying down. But when this stage of belligerence has been reached, unless the goat has a really painful excuse for her behaviour, it is better to abandon mechanical methods of control and resort to psychological methods.

A great deal of trouble at milking time can be avoided if goatlings are accustomed to having their udders handled and massaged long before they are due to kid. The first step in milking a nervous and fidgety goat should always be to stroke and massage the udder until she quiets down: talk to her as you do so, and keep on talking (or singing) as you start milking.

Another great aid to milking without tears or spilt milk is to let the kid pioneer the job for the first four days of lactation, which are the most painful and nerve-racking for the new milker. Most seasoned goat-keepers prefer to put their goats on to a raised platform to be milked; the beginner may prefer to keep the goat on the floor where he can exercise firmer control; but, with the goat standing at a higher level than her milker, there is less disturbance of her coat, and the milk is cleaner; and latterly, when we squat down to goat level, we all tend to creak.

Goats are generally easier to milk than cows; and the worst of them, though tiresome, cannot do any hurt to anyone. The problems created by the difficult milker have been treated in some detail because this is an aspect of reality which goat literature usually tends to gloss over.

Machine milking

Milking parlours for goats, with all the necessary equipment, are readily available from several manufacturers, one of which, Alfa-Laval, has produced an excellent booklet, 'Machine Milking of Dairy Goats' (see Bibliography).

The milk is drawn out of the udder by a vacuum; it is alternate applications of air and vacuum caused by a pulsator which make the soft liner, enclosing the teat, open and close resembling the sucking action of the kid. The vacuum is created by an electric or petrol motor, driving a pump. This must be big enough for the system and needs regular oiling and careful siting. The vacuum system also has a trap to intercept any liquids, a regulator, gauge, pipelines, valves and taps. The vacuum level has to be set for goats, at a slightly lower level than for cows.

There is also the pulsator system which alternately connects the space between the liner and its rigid outer shell to air and vacuum. The pulsation rate has to be set for goats at about 90/minute. The ratio between the massage phase when the liner closes on to the teat, and the milking phase when the vacuum is in operation, also has to be set.

The cluster assembly – teat cups, liners and pulsation tubes – is another requirement. The system to collect the milk – a bucket and cover, or a pipeline system, with or without recorder jars, where the milk goes into a receiver vessel and then is pumped into a storage tank – is a completely closed unit. Bucket systems are cleaned manually, pipeline systems usually automatically with a clean-in-place system.

A milking parlour is usually constructed to enable the goats being milked to be held in place, maybe by a yoke, through which they put their heads in order to eat concentrates while being milked. The goats should stand up on a platform at a convenient height for the operator to put the clusters on. Often there are twelve goats and six milking points. Ancillary rooms for the cooling tank, the machinery and the feed, will be required.

The entire set-up will be chosen depending on the number of goats to be milked. Parlour design is very important to the speed and efficiency of milking – advice should be obtained from as many sources as possible in the planning stage.

Differences between goats' and cows' milk

The average size of the fat globules is smaller in goats' milk and they do not clump together so much, so fat has less tendency to rise. Goats' milk makes a much softer curd in the stomach, crucial to its digestion in babies and the sick. Fats contain fatty acids with different numbers of carbon atoms. Goats' milk is richer than cows' milk in C_6, caproic acid, C_8, caprylic acid and C_{10}, capric acid. Unfortunately the milk fat breaks down easily to leave these free acids, which cause the 'billy' taint, but these short chain acids make goats' cream easier to digest.

Protein

Alpha-s-I casein, which predominates in cows' milk, is almost absent in goats' milk. This may be why goats' milk gives a lower yield of cheese.

Goats' cream

This is a relatively easy form of produce both to make and to market. If the skim milk can be fed to remunerative stock – calves, pigs, orphan lambs, pedigree pups and kittens, mink, pedigree chicks, etc. – the cream can be sold to compete in price with cows' cream. It will always compete in quality: children prefer goats' cream because it is more digestible; housewives prefer it because it whips to greater bulk. In flavour it is not readily distinguishable from cows' cream; but goats' cream is dead white in colour.

The prime requirements of the cream producer are: (a) a remunerative outlet for skim milk; (b) a strain of goats with high butterfat milk; (c) a goat diet rich in fibre. For mass production, it is also essential to have a good separator: not all separators are satisfactory for separating goats' milk. Alfa-Laval, Gascoigne, and Lister separators give good results; the farm (over 2-gallon [9·1 litres]) models give better results than the smaller models. A fine setting of the cream screw is necessary in all models, for the fat globules of goats' milk are small, and slow to respond to centrifugal force. If in doubt whether a given separator or screw setting is giving satisfactory cream extraction, run the separator at full speed, and let the milk tap run at half or quarter its capacity: if you get a better cream output per gallon in this way, your cream-screw setting is too wide, or your separator is not readily adaptable to goats' milk. A cream-screw setting or a separator which gives good results at the start of lactation may give poor results later on, for the size of the fat globules in the milk decreases as lactation advances.

The type of cream produced must depend on the taste and nature of the market; the creams which are subjected to much heat treatment before or after separation do not whip satisfactorily, and when they sour they produce foul flavours. The creams subjected to little heat treatment do not keep so well; but, if cleanly produced, they sour pleasantly, and for whipping they are superlative.

Thin cream

Run the milk through the separator while still warm. If the milk from more than one milking is used, it should be pasteurized when new by holding it to 180°F (82·2°C) for 30 seconds, then cooled quickly and stored in a cool place until required. Before being put through the separator, the milk should again be raised to 180°F and rapidly cooled to 100°F (37·8°C) before being separated. At least part of the milk used for thin cream production should be 'warm from the goat' when put through the separator; otherwise, the cream will not sour pleasantly.

An extraction rate of 1–1⅛ pints (568–639ml) from 1 gallon (4·5 litres) of milk will give cream with a fat content comparable to that of most of the cows' cream on the market. In some districts this product is termed 'ream', the term 'cream' being reserved for:

Double cream

This is produced in the same way, but with a cream-screw setting to produce ¾ pint (0·45 litres) per gallon (4·5 litres) of milk.

Thick granulated cream

Proceed as for double cream, then immediately cook in a bain-marie or double saucepan for 20 minutes at 180–200°F (82·2–93·3°C). Cool rapidly, and place in a refrigerator for 12 hours at least. This form keeps extremely well, and is ideal for eating with fruit; when whipped, it forms butter very rapidly, but the butter has poor keeping qualities. Still the best form of cream for long-distance marketing; it would stand summertime transport from the Hebrides to London.

Devonshire cream

Cool fresh-drawn milk quickly, and leave it to stand in shallow pans in a cool place for 12–24 hours, according to air temperature. Then place the pans on slow heat until the surface cream starts to wrinkle and crack; do not let it boil. Cool again quickly, and skim with a skimmer or saucer. The resultant cream can then be sold for quick consumption, or heated in a bain-marie or double saucepan to 180°F (82·2°C) for 30 seconds and re-cooled, for better keeping qualities.

Butter

Butter is not easily marketed at an economic price, owing to the competition of imported cows' butter and the operation of the food subsidies. In large cities where there is a significant foreign population of Greek or Cypriot extraction, and specialized provision stores catering for them, a regular market for goats' butter may be developed – goats' butter is held in special reverence by these people, and traditionally occupies an honoured position on the menu for festal occasions.

But well-produced goats' butter is such a superlative product that, once it has been savoured, the best of cows' butter becomes, in comparison, uninteresting grease. The goatkeeping household will inevitably demand it.

For occasional production, the flavour of the butter can usually be safely left to natural souring organisms, provided there is a high standard of dairy hygiene throughout. But for regular production, it is necessary to control the souring organisms and use a dairy butter starter.

Natural sour butter

Separate the milk as for double cream, adding fresh batches of cream daily, or allowing the cream to stand until it smells slightly acid but does not taste sour. Take the temperature of the cream with a dairy thermometer (a cheap implement that saves an immense amount of work in butter making). Bring the temperature of the cream to about 57°F (13·9°C) by standing the can in warm or cold water. If the air temperature is above 60°F (15·6°C), 56°F (13·3°C) is warm enough for the cream; if the air temperature is below 50°F (10°C) bring the cream to 58 or 59°F (14·4 or 15°C). Put the cream in the churn, and churn for about 15–20 minutes – an egg whisk and a bowl make an adequate churn for small quantities. As soon as granules appear in the thickened cream – the noise of churning will change its note at this stage to a 'hollower' sound – add 1 teacup of water for each 1¼ pints (710ml) of cream. If the air is warm (above 60°F), the water should be cold; if the air is cold, the water should be at about 60°F. Continue churning till the butter granules are the size of peas; then strain off the buttermilk, and wash the butter in successive rinses of cold water until the water runs clear. Then, and only then, gather the butter granules into one lump, and start working the lump in further rinses of cold water until no further buttermilk is exuded. Lastly take out the butter and work it on a board with wooden pats (Scotch hands) which have been soaked in brine, to express all the water. Make up into shapes, and leave to stand 12 hours in a cool dairy.

If preferred 'salt', dairy salt should be added to the last rinse of the butter granules, and the butter granules left to stand in this for 15–30 minutes.

If difficulty is experienced in getting the butter to break, a not uncommon trouble when the fat globules become very small in late lactation, use the alternative method.

Salted-cream butter

This method is specially designed to ensure the production of well-flavoured butter from small quantities of cream accumulated over a period of a week. The essential factor is that the daily accumulation of cream should be reasonably uniform – say a pint (0·6 litres) a day.

Extract the cream as for natural-sour butter. To the first batch of cream, add 9oz (255g) of dairy salt per lb (454g) of cream, or 11oz (312g) per pint (0·6 litres); store in a cool, dust-free place with good ventilation; each day stir in another batch of approximately the same volume as the first, but no more salt. At the end of the week, churn as for natural-sour butter. In cool conditions, the accumulation can continue for a longer period, but the amount of salt used must be adjusted so that the amount of salt added to the first batch is 8 per cent of the weight of the final accumulation of cream. The principle of the system is that, under normal dairy conditions, cream will not sour if the salt content is over 8 per cent.

Starter sour butter

Separate the milk as for double cream. Place the cream in a double saucepan or bain-marie, and heat to 180–200°F (82·2–93·3°C) for 20 minutes. Cool rapidly, and place in a refrigerator for 12 hours. If no refrigerator is available, heat to 180°F for 40 minutes and cool rapidly, as low as water temperature will take it. Then warm the cream to 50°F (10°C) and add dairy starter. Cover, and leave to stand until the cream smells acid. If a full-flavoured butter is liked, the cream may be left until it tastes slightly sour (but only when dairy starter is used). Then proceed as in previous method.

Cheese

Cheese is a goat product for which there is the most constant demand, if not always the largest market. From a nutritional view, cheese is superior to any type of meat as a source of protein; and, in this country, it will always be cheaper to produce cheese than its protein equivalent in meat. From a gastronomic standpoint, cheese made in part or whole from goats' milk provides a most effective relief from the dull monotony of factory-made Cheddar and Danish Blues. There is undoubtedly an opening for enterprising cheesemakers in this country to produce cheeses of Continental quality, for which upwards of 25 per cent of goats' milk is either essential or desirable.

But the production of any kind of cheese on a commercial scale requires special skill, and a great deal of technical knowledge in so far as the hard cheeses are concerned. The character, texture and flavour of a hard cheese depend upon precise control of temperature and acidity; such control is beyond the capacity of the amateur, and the professional is unlikely to benefit from any advice the present writer can offer.

There remain a few recipes for hard cheese, semi-hard cheese and soft cheeses, which do not require a great deal of skill in the control of acidity and temperature, and which are suitable for cheesemaking with small quantities of milk. Some samples are given below. The quality of the resulting product will in most cases depend on the type of organisms responsible for 'ripening' the cheese milk. The goatkeeper whose milk normally and regularly sours to a pleasant flavour can risk the use of naturally soured milk as a ripening agent, for occasional cheese production. But anyone seriously intending to produce cheese regularly for the market is advised to cut the risks and improve the uniformity of the product by using a dairy 'starter'.

Management of the starter: the starter is obtainable from dairy supply companies, and will not cost much more than £1·50 per dried packet. It is simply a bacterial culture of suitable ripening organisms. Once activated, it will keep only for a short length of time, but it can be propagated as follows. Simmer fresh milk for 30 minutes, cool to 80°F (26·7°C), add 2 tablespoonfuls of starter per pint (0·6 litres) of milk and keep at 70–80°F (21·1–26·7°C) till the following day, when the milk will have been ripened to a smooth creamy consistency, and constitutes a new supply of starter. The process may be continued for months at a stretch provided the job is done in clean, dust-free surroundings, and the culture is kept at the correct temperature during the incubating process. Eventually the starter will lose its vigour and flavour, and a new starter will be required.

In recent years a 'direct vat inoculum' dried starter has been available, which is much simpler to use, as it is simply sprinkled on the warm milk and stirred in. The 'direct vat' starters can be added to the milk without activation, but are not suitable for carrying over from one batch to the next.

If starter is not available for use in any of the recipes given below, twice the recommended quantities of buttermilk or naturally soured milk, or ten times the quantities of milk held over from the previous morning, may be used instead of starter – with the risks of failure implicit in the use of uncontrolled ripening organisms.

Rennet: only cheese-making rennet of a reputable brand is suitable. The grocer's junket rennet is quite useless.

Salt: use dairy salt only.

The cheese milk: with a few specified exceptions, fresh whole milk is required for all the recipes given. A high standard of dairy hygiene is required to produce good cheese milk. Spoilage bacteria cause foul flavours. If there is any doubt about the quality of the milk, it is better to drink it than use it for cheesemaking. Pasteurization of the milk alters the character of hard cheese, and is responsible for most of the unpleasant and soapy characteristics of the worst factory-made cheese. However, where milk has to be held over from day to day, it is preferable to pasteurize than to risk the milk's becoming over-ripe. When pasteurizing for cheesemaking hold the milk at 145°F (62·8°C) – no higher – for 30 minutes.

The first four recipes given are suitable for goats' milk or for a mixture of cows' milk and goats' milk. The remaining recipes are for 100 per cent goats' milk.

Crofter cheese

A semi-hard cheese of mild flavour maturing in about four weeks and yielding about 18oz (510g) per gallon (4·5 litres); requires about 2 gallons (9·1 litres) of milk. This cheese is of better keeping quality and flavour than that produced by the majority of simple cheesemaking recipes.

Equipment: a tub or pail, a long bread knife, two 15in (38cm) squares of smooth cotton cloth, a perforated steel mould 5¾in high × 4in (146 × 102mm) diameter, 3 weights of 28lb (12·7kg) each – or a 6ft (1·8m) plank with one end hinged to the wall and a 28lb weight on the other (this exerts a pressure of 28lb at the weighted end, 56lb (25·4kg) at 3ft (0·9m) from the weighted end, and 84lb (38·1kg) at 4ft (1·25m) from the weighted end, when the plank is kept horizontal.

Procedure: strain and cool 1¼ gallons (5·7 litres) of evening milk. In the morning, warm it to 70°F (21·1°C), and add 4 tablespoonfuls of starter. Strain into it 1¼ gallons of morning milk, adjust the temperature of the mixture to 86°F (30°C) and maintain this temperature. Then, 1½ hours after adding the starter, add 1 teaspoonful of rennet diluted in 6 teaspoonfuls of cold, clean water. Cover and keep warm for 40 minutes. Then test the curd: insert the index finger obliquely into the curd, run the thumbnail along the surface of the curd above the inserted finger, then raise the finger. If the curd is ready for cutting, it will break cleanly and no curd will adhere to the finger; if it is not ready, test again at 2-minute intervals until it is.

When the curd is ready, cut it with the bread knife into vertical slices about ¼in (6mm) thick; then cut the slices across to make ¼-in square columns; gently turn the columns on their sides with the hand or the flat of the knife and slice them up into ¼in cubes. The job must be thoroughly done and takes time – about 15 minutes. The remaining lumps of curd which you cannot get at any other way should be sliced against the hand. Stir gently for 5–10 minutes.

Now remove 1 pint (0·6 litres) of whey in an enamel or other jug, and place it in hand-hot water, leaving it there until the temperature of the whey in the jug is 105°F (40·6°C). Return the whey to the curd, remove another pint and repeat the process, removing up to 1 quart (1·1 litres) of whey in the later stages, until the temperature of the whole is 94°F (34·4°C). This should take about 45 minutes; a little longer is a lot better than a little shorter. Stir gently while this heating process is going on and, after it is completed, stir for 15 minutes. Then let the curd settle, cover the container and leave for 20 minutes.

Now test the curd for draining. Take a small piece of curd, squeeze it in the hand, and press it briefly but firmly against the surface of a hot, clean poker (or electric iron). Draw the curd gently away from the poker, and if fine threads barely ⅛in (3mm) long are formed, the curd may be drained immediately. Strain the whey off into another vessel through a straining cloth.

Bundle the curd in the straining cloth, put a warm pudding basin upside down in the cheesemaking bucket or tub, and place the bag of curd on top of it. Cover the bucket. At 15-minute intervals during the next hour, open the bundle of curd, tear the lump of curd into fist-sized pieces and replace them in the bundle, turning cold faces to the inside of the mass. A 7lb (3·2kg) weight placed on top of the bundle will assist matting.

At the end of the hour, test the curd against the hot poker again. When the threads so formed reach ½in (13mm) in length, the curd is ready for salting. Break it up into walnut-sized pieces, add a good tablespoonful of dairy salt, and stir the salt into the curd for a minute. Bundle up the curd again, leave it for 5 minutes, and give it a final stir before packing into the mould. Do not let the curd get cold. The hot poker test can be replaced by measuring the acidity of the whey.

Warm the mould in tepid water, and press the curd into it by handfuls, pressing just hard enough to cause a little whey to ooze out. Finish by heaping the centre, and placing the lid in position. Apply 28lb (12·7kg) pressure and leave for 1 hour. Then turn the cheese in the mould, inserting a fresh, dry, smooth cloth as a liner, apply 56lb (25·4kg) pressure and leave till next morning. Turn the cheese again the following morning, renew the cloth, and apply 84lb (34·1kg) pressure. The same evening or the following morning, the cheese should be bandaged.

Rub the cheese with lard, and apply cheesecloth caps to the two ends. Then sew a wide over-all bandage round the cheese, trapping the cap at either end.

Keep the cheese for 4 weeks at least in warm weather, a little longer in cold weather, in cool room temperature (55° to 60°F) (12·8 to 15·6°C). Turn it daily; if mould starts to form, wipe it away with a cloth soaked in brine.

The times given here should be right for a cheese made with mixed milks; when the proportion of goats' milk is higher than 50 per cent, the times given may be slightly shortened – but the heating of the curd must not be speeded, especially in the earlier stages.

Beginners at cheesemaking will find a clockwork 'pinging' timer a great aid.

A proper cheese-press such as the 'Wheeler' is preferable, and easier, than using weights. Also coating the cheese with molten cheese-wax is easier than sewing bandage over lard, and prevents mould from growing. Cloths and equipment should be boiled before use.

(This recipe is the most ambitious given here, and is the most suitable for production for a regular market.)

Little Dutch type

A softer and more open-textured cheese than the Crofter, with less flavour. The quality of this cheese is even more dependent on the quality of the cheese milk.

Equipment: as for the Crofter cheese.

Procedure: prepare the cheese milk as for the Crofter, but raise the temperature of the mixed milks to 90°F (32·2°C) before renneting. Rennet the milk 30 minutes after adding the starter, using 1¼ teaspoonfuls to 1 wineglassful of cold water. Otherwise proceed as for the Crofter; but the curd will require a little longer before it is ready to be cut. In cooking the curd, raise the temperature to 105°F (40·6°C) in 1 hour; the whey should not be heated above 100°F (37·8°C) for the first 30 minutes, nor above 110° to 115°F (43·3 to 46·1°C) for the second 30 minutes. Once the temperature of 105°F is reached, stir the curd for 15 minutes, let it settle, and strain off the whey immediately. As we do not want to bundle the curd in this case, it is better to bale off some of the whey and decant the rest. Immediately add 2 heaped dessertspoonfuls and 1 level dessertspoonful of dairy salt and a pinch of saltpetre; stir in as for the Crofter. The cheese mould should be at the same temperature as the curd, which is packed into the mould as quickly as may be after the mould is removed from the whey. Pressing, turning and bandaging is carried out as for the Crofter.

Typical 'little Dutch' moulds can be used for this cheese and, provided a good starter is used, the cheese milk can be pasteurized without detracting from the character of the cheese in this case. Unless the cheese milk conforms to the highest standards, it is safer to pasteurize.

Wensleydale cheese (*Lady Redesdale's recipe*)

Use 2 gallons (9·1 litres) of milk for 2lb (0·9kg) of cheese. Heat milk to 82°F (27·8°C). Add 1 teaspoonful of rennet mixed with 5 teaspoonfuls cold water, and stir for ½ minute. Leave 1 hour.

Cut the curds into ½in (13mm) cubes, as for Crofter cheese. Stir gently for 5 minutes with the hand.

Raise the temperature of the curds to 86–90°F (30–30·2°C) in the course of 20–30 minutes, by standing the curd bucket in a larger vessel of hot water and stirring constantly. When the cubes of curd are firm and 'shotty' leave to stand for 10 minutes. Drain off the whey, put the curds in a cloth, and squeeze out the whey. Lay a 10lb (4·5kg) weight on the bundle of curd, and leave for 10 minutes. Break the curd and turn the cold faces inwards, bundle up, and replace weight for a further 10 minutes; repeat this last process. Then break up the curds by hand, and salt to taste. Leaving the curds in the cloth, place cloth and curds in a mould, fold the top of the cloth flat, and apply a 10-lb weight. Turn daily for 3 days, replacing the weight each time. After 3 days, remove the cheese from the mould, and place it on a shelf in a cool, airy place, turning it daily. It will be ripe in 3–4 weeks.

Brie type (*Recipe from a long-established Anglo-Nubian goat dairy in Eire*)

Bring 2 gallons (9.1 litres) of perfectly fresh, clean milk, preferably warm from the goat, to 83°F (28·3°C). Mix 2ml cheese rennet in 10ml of water. Stir into the milk. Leave to stand for 2 hours.

Ladle the curd into hoops, 10in (25cm) in diameter and 3in (8cm) high, standing on a straw mat on a draining-board. Allow to drain for 24 hours, or until the curd is firm enough to hold its shape without the hoop; sprinkle salt on the upper surface of the cheese, and leave for another 24 hours. Turn the cheese on to a fresh straw mat, and rub salt into the second side. In a few hours turn again and leave to ripen.

Turn the cheese each day in a dry well-ventilated atmosphere. In about 8 days, moulds begin to grow on the surface; the cheese should then be transferred to a dark cellar with little ventilation and with a temperature around 55°F (12·8°C). The ripening is brought about by the moulds, which grow on the surface, diffusing enzymes into the cheese. If conditions are not right for the moulds, the cheese will spoil; mould growth is the key to the operation. The cheese should ripen in 2–4 weeks, by which time it should have a semi-liquid or waxy centre.

This recipe could present considerable health risks. A proper inoculating mould should be used, and the milk pasteurized first. Low-acid cheeses of this type are ideal media for growth of *Listeria monocytogenes*, the causal organism of listeriosis. The dairy in Eire was fortunate in having the right bacteria growing in its cellar!

Pont l'Evêque type

This resembles the distinctive Normandy cheese which is made from cows' milk. Yield: 2 cheeses about 5in (13cm) square and 2in (5cm) deep, weight 15oz (425g) from 1¾ gallons (7·9 litres) of milk.

Equipment: 2½-gallon (11·4 litres) pail for the cheese milk, a sharp-edged skimmer. A large tub or dairy sink, in which the pail can be immersed (not necessary in warm weather); a bread knife for cutting the curd; 2yd of 36in (1·8m of 0·9m) cheesecloth; straw or bamboo mats, 1½in × 8in (38mm × 20cm); draining boards 14in × 8in (35 × 20cm) – preferably 4 mats and 4 draining boards for each pair of cheeses, but two will do. Two steel moulds, square with rounded corners, 4⅞in square × 2¼in (124 × 57mm) deep; a draining rack about 30in × 12in (76 × 30cm) with ⅜in (9mm) spars set about ½in (13mm) apart. (All standard cheese-making equipment obtainable from dairy suppliers.)

Procedure: strain the morning's milk – 1¾ gallons (7·9 litres) – into the cheese pail; adjust the temperature to 92°F (33·3°C) and place the pail in the tub or sink half-filled with water at 94°F (34·4°C). Or work in a warm draughtless room during warm weather. Add 2 drops of starter, stirring in each in turn. Dilute ½ teaspoonful rennet in ½ wineglassful of cold water, and add this to the cheese milk; stir for 5 minutes. Cover the pail with a cloth. Leave for 45 minutes; but in cold weather, remove a little of the water in which the pail is standing every 15 minutes, and replace with water at about 94°F. The temperature of the water may be allowed to drop, but no quicker than it would drop on a good spring day.

After the 45 minutes, test the curd for cutting as in the Crofter cheese; it may require up to 1 hour before it is ready; if it takes any longer, the cheese milk has not been kept sufficiently warm. When the curd is ready, cut it into 1in (2·5cm) square columns, and then cut diagonally through the squares, once only, to give triangular columns with 1in sides. Leave for 10 minutes.

Then, with a skimmer, ladle the curds in ¾in (19mm) slices into a previously scalded straining cloth spread over the sparred rack. Fold the ends of the cloth over the curd, cover with another dry cloth, and leave for 30 minutes.

Open the cloth, and quickly cut the curd into 3in (8cm) squares. Fold back the cloth and re-cover with the dry cloth. Leave for another 30 minutes.

Take one corner of the folded cloth and bind it tightly round the other three, bundling the curd slightly in the process. Re-cover with the dry cloth. Leave for 15 minutes. Tighten the bundle and turn it upside down, with the knot underneath. Leave for another 15 minutes.

Place the 2 moulds together on a scalded straw mat, and the straw mat on a draining board. Then open up the curd, breaking it into strips and small pieces as you transfer it into the moulds. The curd lying against the side of the mould should be firmly pressed against the side; as each layer of curd is laid, sprinkle it with dairy salt, using 1 heaped tablespoonful of salt to each cheese.

When the moulds are filled, cover them with a scalded straw mat, with a draining board on top of that. Now comes the beginner's anxious moment. Grip the two boards between the fingers of both hands, and turn them over so that the bottom of the cheese is now uppermost. Repeat this process at 10-minute intervals for the next hour, and once more before you pack up for the night.

Leave the cheeses in a cool, airy room, turning them in the same way each day; it is preferable to change the top mat and draining board for a fresh and scalded pair if the weather is warm and muggy. After 4 days from the making of the cheese, remove the moulds. Smooth the sides of the cheese with a knife to fill in any crevices, lay the cheeses on a dry muslin and turn them each day until the surface is quite dry – about a fortnight if conditions are favourable.

Now wrap each cheese in greaseproof paper and wrap each pair of cheeses, or preferably each 4 cheeses if you have them, in a cheese cloth, scalded and dried. Place the bundles so that the cheeses are lying on their narrow sides, on a latticed shelf or baking tray in a cool, airy room. Every day, turn the bundles round so that the cheeses lie on another part of their sides; every alternate day open the cloth bundles and turn the cheeses so that the side of the cheese facing inwards now faces outwards. If there are 4 cheeses in the bundle, move the inside ones to the outside.

Ten days after the cheeses are wrapped, they should be unwrapped completely, and any mould growth on them wiped off with a cloth soaked in boiled and cooled brine. A cheese with much mould growth on it should be dipped in the brine solution and allowed to dry before being rewrapped in clean paper. If, while turning, any mould is noticed on a cheese at an earlier date, it should be treated in the same way.

The cheese should be ripe 2–4 weeks after wrapping, and is best eaten within the following fortnight – hence the reason why the cheese is not widely imported.

The Mont d'Or

France's most popular 100-per-cent goats'-milk cheese is made in a very similar way, but by people who apparently attach no particular value to a 50-hour week. I have been unable to obtain a detailed recipe, but the following may provide early risers with a basis for experiment.

To 2 gallons (9·1 litres) of fresh clean milk at 82°F (27·8°C) is added 1 drop of starter and 4 drops of rennet diluted in 2 teaspoonfuls of water – each being thoroughly and separately stirred in. The cheese milk is kept in a warm place, and the curd is ready for cutting in about 8 hours, or a trifle longer. The curd is then cut as for the Pont L'Evêque, and is strained in a 9in (23cm) long, conical strainer for 4 hours, no attempt being made to conserve the heat of the curd. Thereafter, the curd is packed into circular moulds a little over 4in (10cm) high and about 3in (8cm) in diameter, with 2 heaped tablespoonfuls of salt sprinkled between the layers. A scalded condensed-milk tin with the rims removed would seem to provide an experimental mould. The moulds are placed on and covered by the same arrangement of straw mats and draining board as the Pont l'Evêque, and the cheeses are frequently turned in the same way. After 3–4 days, when the cheese can hold its shape without the aid of the mould, the mould is removed, and the cheese is rubbed with salt and dried for a further few days on clean muslin laid across a slatted rack or baking tray. When sufficiently firm, each cheese is completely enveloped in a cheese bandage or wicker basket, and hung up to dry in a cool, airy room. The cheese ripens about 10–15 days after leaving the mould in the summer of central France, probably a little longer in our climate.

With a mould of this height, a few elastic bands strategically stretched between the draining boards would appear advisable.

Ross-shire Crowdy butter

This is not, strictly speaking, a cheese as we know it; but it is a modernized version of an eighteenth-century Highland recipe for a curd-and-butter mixture, based on the use of goats'-milk curd and cows'-milk butter. It produces a mellow, highly flavoured product somewhat akin to some of the Flemish cheeses.

To 1 gallon (4·5 litres) of fresh milk at 90°F (32·3°C) add ½ pint (0·3 litres) of starter, stir thoroughly, and leave in a warm place for 6–12 hours, when the whole will have become a smooth, creamy curd. Heat to 110°F (43·3°C) by standing it in hot water and stirring; maintain this temperature for 30 minutes–1 hour, stirring continuously until the curd hardens sufficiently to drain off the whey. When the whey has been drained, the curd is washed with sufficient cold water to reduce the temperature to 70°F (21·1°C) and strained through a cloth. The cloth containing the curd has its four corners tied together, and is hung up to drip for 1–2 hours, when the curd should be of firm pasty consistency.

The curd is then thoroughly mixed with 1 heaped tablespoonful of salt and ¾lb (340g) of butter. A scalded wooden box 7in long, 1½ wide and 4in deep (178 × 38 × 102mm) is lined with two strips of cheesecloth that have been scalded in boiling brine, the one strip being 1½in wide and 23in (58cm) long, the other 7in wide and 12in long (18 × 30cm). The curd and butter mixture is rammed into the box tightly, with particular attention to filling the bottom corners, and the loose ends of the cheesecloth are turned over to cover the top of the mixture.

Any space remaining at the top of the box is packed with salt, a lid is fixed and tied firmly down, and the box is stored in a cool airy place for 4–6 weeks: 4 weeks is sufficient for the first trial.

If this recipe is cut short after the salt is added to the curd, without the butter, you have the recipe for crowdy or cottage cheese, which must be eaten within a few days of making. It is, incidentally, the better for having the butter worked in, even if it is to be eaten fresh. Skim milk can be, and frequently was, used for this recipe.

Norwegian whey cheese

This is a useful method of using up whey. Evaporate on low heat, stirring often until creamy, and thereafter continuously until pasty in consistency. Spoon into greased bowls to cool; tip out. It is greyish-brown, with a concentrated sweet-salt-sour flavour, and very nutritious; the market is small, but enthusiastic.

Smallholder cheese

To 2 pints (1·1 litres) of fresh warm milk add ½ teaspoonful of rennet and stir well. Let stand for 12 hours, then cut the curd into pieces of uniform walnut size. Transfer the cut curd to cheesecloth and hang it up to drain for 24 hours. Then add a pinch of salt to the curd and place it in a mould stood on a small straw mat. Place another mat on top and slightly weight down.

After one day turn the mould right over, use clean mats and weight from the other side to compress the curd.

After a further day the cheese is ready for use.

Lactic acid curds

Use whole, skim or separated milk. Heat to boiling point and make acid in one of the following ways:

(a) 2 tablespoonfuls of vinegar to every 3 pints (1·7 litres) of milk.
(b) 2 tablespoonfuls of lemon juice to every 3 pints of milk.
(c) ½ teaspoonful of tartaric or lactic acid dissolved in a cupful of hot water and added to 3 pints of milk.

Stir in the acidifying agent. The milk protein will curdle and the whey separate immediately. Strain through a cloth or fine nylon sieve.

This curd will have little flavour and can be used in sweet or savoury dishes, such as the basis for Yorkshire curd/cheese cakes; or add chopped herbs (chives, lemon balm, etc.) and use as a spread.

It can also be pressed under a 4lb (1·8kg) weight, then cut into cubes and used in curries.

Mrs Leueen Hill of Redruth, Cornwall, kindly supplied the two previous recipes.

Yoghurt

First made in the Balkans, and with goats' milk, which results in a slightly less acid produce. The essential element is rapid conversion into preservative lactic acid of the lactose (milk sugar) by a pure culture of bacteria which leaves all other constituents unaffected. Whereas other preservative treatments impair flavour and nutritive qualities, this process preserves and enhances protein content, and leaves fat content and vitamin potency intact.

To make yoghurt: bring 1 pint (0·6 litres) milk to the boil and simmer for 1 minute. Leave milk to cool until it reaches 98·4° to 100°F (37 to 40°C). Put 1 generous dessertspoon of live yoghurt into a bowl and slowly add the cooled milk, stirring thoroughly so that the two are well mixed. Transfer the contents to a wide-mouthed vacuum flask, screw on the lid and leave the flask to stand for about 8 hours. This is an extremely reliable method, but in the absence of a suitable flask the bowl can be covered tightly with a plastic bag, wrapped in newspaper or blanket or towelling and put in a warm, draught-free place (an airing cupboard or the back of a solid fuel stove, for instance) and again left for 8 hours, or a little longer if necessary.

Milk must be initially clean and kept free of contamination by other bacteria.

Various strains of yoghurt bacteria are available from health-food shops and specialist suppliers; each imparts a slightly different flavour, and demands a slightly different treatment, which will be specified. For goats' milk, the Bulgarian strain is best.

True yoghurt is an almost pure culture of the specific bacteria, and can be used as a 'starter' for further batches. Pseudo-yoghurt, acidified milk totally without live yoghurt bacteria, is easy to produce; easiest of all if lactic acid from the chemist is added directly to the fresh milk seasoned to taste. Its flavour is indifferent; and if you try to incubate the sterile pseudo-yoghurt with fresh milk, it just curdles deceptively.

CHAPTER TWELVE

The Universal Foster Mother

To the supreme honour of a place in the heavens, among the signs of the Zodiac, the Greeks elevated three of their domesticated animals: the Bull who drew their ploughs; the fleecy Ram who clothed them; and Capricorn the Goat. The name of the goat who earned this honour for her species was Almalactea 'Foster milk'. Her constellation still brightens the twentieth-century sky.

While relatively few of the newly born of other species can be satisfactorily reared on cows' milk, however modified, there is probably not a single species among the larger land mammals the young of which will not thrive on suitably adjusted goats' milk.

The purpose of the present chapter is to consider the adaptation of this highly digestible food to the peculiar needs of the various species of farm livestock, and to sugest how Almalactea can turn an honest penny for the twentieth-century farmer.

It is generally accepted that the composition of the milk of each species is ideally adapted to the growth pattern of the young of the species, and that any other milk composition will represent a departure from the ideal.

In other words, goats' milk will be entirely suitable for youngsters of other species which grow at approximately the same rate as the kid: for those that grow faster, it will prove too weak; for those that grow more slowly, it will prove too strong.

It is perfectly true that members of the medical profession are, for the most part, convinced that modified cows' milk, which is designed to suit the fast growth rate of the calf, is perfectly suitable for feeding the slow-growing human infant. It is perhaps sufficient to point out that farmers are concerned with the economic life history of their stock, while doctors are in practice primarily concerned with resolving immediate problems, and are seldom able to test the long-term effects of their recommendations. There is a massive collection of scientific evidence to show that all young farm stock, if fed a diet too rich in digestible protein, are liable to mineral-deficiency disease sooner or later.

No one has any doubt that all youngsters must be fed a diet

Table 17 Comparative composition of the milk of various species

Species	*Water (%)*	*Fat (%)*	*Sugar (%)*	*Casein (%)*	*Other protein (%)*	*Ash (%)*
Goat	86·2	4·5	4·08	2·47	0·43	0·79
Cow	87·3	3·67	4·78	2·86	0·56	0·73
Sheep	79·46	8·63	4·28	5·23	1·45	0·97
Mare	89·8	1·17	6·89	1·27	0·75	0·30
Donkey	89·88	1·5	6·09	0·73	1·31	0·49
Dog	75·44	9·57	3·09	6·10	5·05	0·73
Cat	81·63	4·49	4·79	3·72	3·30	0·58
Pig	83·23	4·5	4·2	7·3		0·77
Woman	87·4	3·0	6·5	0·04	0·7	0·25

sufficiently rich to maintain normal ratc, if thcy arc to bc hcalthily reared.

Table 17 shows the comparative composition of the milk of the various species of domesticated animals, including women (!). Table 18 shows the adaptation of goats' milk which is required to produce a milk as nearly as possible ideal for the rearing of various classes of farm stock. It will be seen that this adaptation is quite satisfactory for calves, lambs and foals; it is not so good for puppies, but valuable orphan pups can almost always be reared on this mixture, which gives them insufficient protein, to the stage at which they can take enough of the milk to provide all the protein they need. For rearing kittens from birth goats' milk is the best substitute for cats' milk available, but will only rear the stronger members of a strong litter. Few people will be unduly concerned at this lapse from perfection.

The details given in Table 18 should suffice to tide the farmer and his orphan over a crisis. But for the goatkeeper who wishes to develop the commercial possibilities of 'foster milk', other considerations are of importance.

The best place for a goats'-milk orphanage is on an unspecialized smallholding. In any reasonably densely populated farming district, there is a reliable if seasonal supply of orphan lambs, orphan piglets and piglets surplus to the milking capacity of the sow, with an occasional orphan foal to add interest. A word to the veterinary surgeon, and a little

Table 18 Adjustment of goats' milk for the feeding of the newly born of farm animals

Species	*First ten days*	*Thereafter*
Calf	Whole milk to appetite in four feeds per day.	Whole milk up to 1 gallon (4·5 litres) per day in two to four feeds.
Lamb	Whole milk with 1oz (28g) of thin cream added to each ½pint (0·3 litres) to appetite in four feeds per day.	Whole milk up to 2 pints (1·1 litres) per day for hill lambs, and up to 4 pints (2·3 litres) per day for other breeds, in four feeds per day.
Foals (including donkey foals)	Half–and–half milk and water with 3 level tablespoons lactose (sugar of milk) to each pint of *milk*, or 2½ level tablespoons sugar to each pint of *milk*, and 1 teaspoon lime-water to each pint of *mixture*, in four feeds of up to 1 pint (0·6 litres).	The same, feed to appetite in four feeds a day. At 2 months the proportion of milk can be gradually increased to 75 per cent of mixture.
Pups	1 teaspoon of thin cream in a tablespoon of whole milk at eight feeds per day; feed to appetite.	The same at six feeds till 3 weeks. Then whole milk to appetite in four to six feeds.
Kittens	Whole milk in six to eight feeds per day.	The same, reducing number of feeds.
Piglets	1 dessertspoon of cream in 1 teacup of milk. 2–4oz (57–113g) at six feeds per day. Or 2 dessertspoons Glucodin instead of cream.	Whole milk with 1 tablespoon of cream per pint (0·6 litres) to appetite at four feeds. Trough feed at 3 weeks. Replace Glucodin with brown sugar.

Notes on table:

(1) Colostrum from a newly kidded goat may be given to the new-born calf and the new-born lamb, but not to any other new-born. One feed is sufficient.

(2) All feeds must be fed at blood heat and at regular intervals. But the interval before the last feed at night may be longer than the others. Feeding bottles must be sterilized after each feed.

(3) *Calves* Give 1 dessertspoon of olive oil in the first feed if the calf cannot receive colostrum.

(4) *Lambs* Give 1 teaspoon of olive oil in first feed if the lamb cannot obtain colostrum. Feed milk in a polythene baby's bottle fitted with lamb teat, or in a nylon 'fre-flo' bottle.

(5) *Foals* Give a first feed of 1 dessertspoonful of oil in *whole* milk if the foal cannot receive colostrum. Feed from a wine bottle fitted with calf teat.

(6) *Pups* Add 2 drops of cod-liver oil to the first two feeds. Feed with a disposable plastic hypodermic syringe with no needle. Add 'Sister Laura's Food' for weaklings and miniatures.

(7) *Kittens* Feed with a plastic syringe.

(8) *Piglets* Add 1 teaspoon of cod-liver oil to first two feeds if the piglet has not had colostrum. Feed with a polythene baby's bottle and baby's or soft lamb teat.

market-day advertising, will canalize the supply in the required direction. Orphan lambs can be reared on cows' milk, though they seldom do nearly as well as they do on goats' milk. Consequently, orphan lambs command a fair price. Piglets reared on cows' milk from birth never pay their way, and usually die. So orphan and surplus piglets are cheap, and provide the staple throughput of the goats'-milk orphanage.

The first rule of the orphanage should be: fresh orphans only. There is little hope of making a profit out of any new-born creature that has already had its digestive system and vitality undermined by a diet of cows' milk or cows' milk and water, or by simple starvation. This is especially true of piglets, who are not worth having unless they are fresh from the sow.

Piglets will take to the bottle easily. The weaker of them may benefit from the addition of a little glucose to the feeding recipe given in Table 18; but this should not be continued for more than two or three days. The piglet's main need during the early stages is for warmth. For a regular orphanage business, an electric infra-red heater is worth while. For casual business, a straw bed of good depth, with a few well wrapped hot-water bottles under the surface, will suffice.

Frequent feeding of the piglets during the first three weeks is essential. Six feeds at, say, 7, 10 a.m., 1, 4, 7 and 10-30 p.m. will suffice without greatly disrupting the peace of the farmer. After three weeks, four feeds a day will be sufficient; and the milk can be fed in a trough, with a little taste of meal and some milk-soaked bread added as soon as they have become accustomed to the change. The meal ration is increased as the piglets grow; and, by five weeks old, they will be able to maintain progress if the whole milk is replaced by skim.

From a few days old, the piglets should have room for exercise; if weather and circumstances permit, they will do best with a grass run; otherwise they must have some turfs or soil to supply the iron they need. If there is a good supply of piglets, the goatkeeper is best advised to sell at eight weeks old, when the piglets will weigh about 40lb (18kg). Sold at this stage they should return a good price in proportion to the milk, etc., that has been fed to them. If kept any longer, the return per gallon will drop substantially and a cream market would have to be found to make the enterprise profitable.

These comments are not intended to describe methods of pig rearing – for which the writer is not qualified – but to clarify the main points of contact between goatkeeper and piglet rearing.

Orphan lambs should always be reared on the bottle, and never suckled on the goat. This is in the best interests of both parties. Ewes have smaller and tougher teats than goats, and a lamb is likely to break the skin of the goat's teat with its teeth, and give rise to black garget infection. The goat's grazing habits are very different from those of the sheep, and a goat suckling a lamb will lead it to pastures unsuitable to the lamb, and teach grazing habits inimical to the lamb's digestion.

Calves will not feature in the goat orphanage; there is plenty of cows' milk. But calf rearing can provide a basis for profitable goatkeeping. We are not concerned here with the farmer who keeps goats and uses their milk to rear his replacement heifers, while drawing the subsidized price for the cows' milk the calves would otherwise have consumed. But a goatkeeper may specialize in calf rearing.

The goatkeeper may reasonably expect to rear a slightly better calf than anyone else for two good reasons; the high digestibility of goats' milk will minimize setbacks from digestive troubles, and will practically exclude the normal liability to white scour; the fact that goats and calves share but few internal parasites will afford both parties a measure of protection when sharing a pasture – the calf will eat the larvae of the worms that infest goats, and the goats will eat the larvae of the worms that infest calves, with no ill consequences to either.

Stock rearing of this kind is one of the least profitable forms of farming in these days. If calf rearing is to be really attractive to the goatkeeper, he must be able to rear a calf which cannot be matched by any other method of rearing, and command the premium his supremacy deserves. There is only one way in which this can be done: for a goatkeeper with a herd of Anglo-Nubians or Anglo-Nubian crosses (that is, a herd producing milk with an average butterfat of over 5 per cent) to rear pedigree heifers of a breed giving low-butterfat milk. A

Plate 15 R77 Osory Snow Goose, owned by Gary and Chris Jameson, Mytaba Stud, Leppington, New South Wales, Australia. (D. Mytaba Sulpaedes.) Holder of Australian and world records for 1st Lactation and Senior Doe. 7267lb on 1st lactation. 7713lb on 2nd lactation.

Friesian heifer reared on Anglo-Nubian goats' milk is a better reared beast than any Friesian heifer has a natural right to be. If you put a Friesian calf on a Jersey cow's milk, it will get the butterfat; but it will also get indigestion. Only the goat can do the job better than Nature.

This is not the job for the novice, nor even for the specialist goatkeeper without other experience. The calves worth rearing are worth three figures at birth. But here is an opportunity for a first-class all-round stockman with a good knowledge of goats, and a special knowledge of bringing out cattle, to do an independent job dear to his heart and good for his pocket.

Calves will literally, and properly, go down on their knees for a drink of goats' milk, and may be suckled on a goat with adequate teats. The system has been practised without mishap on a small scale: whether the saving of labour and additional protection from scour provided by direct suckling is worth the risk of injury to the goat's udder, only time and experience can tell.

How the stars of Capricorn will twinkle when the champion dairy cow is reared on goats' milk!

CHAPTER THIRTEEN

Kid-meat

The meat of adult goats is not particularly good eating, and is generally only consumed *faute de mieux* in most countries; in Britain, it is held in particular aversion. Earlier editions of this book contained information about it which has now been left out, as being a lost cause. One worthwhile recipe for goat ham is retained.

Kid-meat, on the other hand, can be a prime meat, and is an entirely different proposition from any other kind of goat meat. The male kids arrive whether we want them or not, and for four days at least the goat produces milk which is of little use for anything except kid-rearing. So we have a 10lb (4·5kg) kid free.

For those unable to slaughter their own animals for home consumption, slaughter-house charges would be too great for a kid so small. It is illegal to sell or supply meat from an animal not slaughtered at a licensed abattoir, but rearing kids to five or six months old on their mothers' or unwanted goats' milk, out at grass, may prove economic.

From a culinary point of view, kid is rather more versatile than, and almost as well flavoured as, lamb; for the modern taste for lean meat, kid meat may well prove superior. The flavour of meat depends to a great degree on the condition of the animal from which it is derived; it is not possible to bring lambs to the condition in which their meat develops its optimum of flavour without at the same time introducing more fat into the meat than the modern housewife cares to see. But a perfectly conditioned kid will present a lean chop.

It is worth pointing out that the price of meat in the butcher's shop is today approximately twice the price that the farmer receives per kg dressed carcass weight. So kid-meat is worth twice as much if you eat it at home as if you sell it to the butcher.

At present in France, the National Institute for Agronomic Research (INRA) is investigating the increased breeding of goats, of which there are now about one million head in France. Goat production until now has been mainly for the production of cheese, but now consumers are interested in kid-meat. Previously heavy kids were considered a by-product of the dairy industry. Suckling kids used to drink their mother's milk for about two weeks before being slaughtered, while weighing just 13 or 15lb (5·9 to 6·8kg). Now breeders tend to feed kids with milk products and to slaughter them aged between 3 and 5 weeks, weighing between 18 and 26lb (8·1 and 11·8kg) or sometimes even more. Researchers have developed feeding techniques for producing heavier animals. At the same time they have tried to rationalize techniques in order to reduce manpower costs. In some European countries, small kids are an Easter delicacy; the same is not at present widely true in the UK.

Kid-meat production brings one new consideration into goatkeeping practice: the necessity to castrate the male kids at birth. This needs to be done if meat kids are to be run with females, and kept through the winter. Otherwise, entire males grow more quickly than castrates, and castrates, particularly of Angora cross-breds, tend to lay down a great deal of fat, which most consumers nowadays do not like on their meat.

By far the simplest and safest method of castration is by the 'Elastrator', a device for applying a specially designed rubber band which cuts off the supply of blood to the scrotum. It has, by law, to be applied before seven days of age. The Farm Animal Welfare council (FAWC) are currently trying to get this method of castration banned in Europe, as they believe it to be painful and therefore cruel.

From a culinary point of view, kid-meat is of three types. Month-old kid, a white and rather glutinous meat, is like veal, or chicken in the *poussin* stage. It can be used in most veal and *poussin* recipes, but is best in the more highly flavoured ones. Three- to four-month kid can be treated as spring lamb, which is a singularly happy thought in days when there are very few spring lambs to be treated; the better restaurants and hotels can be very grateful for this type of meat. Six- to nine-month kid can be larded and treated like lamb; but it is at its best when marinated, for then it becomes as well flavoured as prime venison and as tender as prime lamb, and can be used in recipes designed for either meat.

The 'ennobled' Boer goat has been developed in South Africa as a meat breed, and a good full-grown male specimen has an impressive bull-like appearance. Individuals reached the UK via breeders in Germany in 1988. The body is white, the head red-brown (Plate 6), and the nature docile. The British Boer Goat Society is affiliated to the BGS.

Some recommended recipes follow.

Roast kid (*Central France*)

A leg of kid of three to four months old, weighing 3–4lb (1·4–1·8kg) is pierced with the point of a carving knife in eight to twelve places where the flesh is thickest. Into half of the gashes place a piece of peeled garlic the size of a split pea; into the rest of the gashes force ¼in (6mm) strips of bacon fat or salt pork.

Place the kid in a roasting tin with ½ pint (0·3 litres) water and 1 teaspoonful of salt; break 1oz (28g) of butter into small pieces, and dot them over the top of the joint.

Roast in the oven at 340°F (175°C), gas mark 3½), allowing 20 minutes per lb (454g).

Remove the joint into a heated dish and keep hot. Scrape the bottom of the roasing tin, add a little water if necessary, strain, and serve this gravy with the meat. Garnish with watercress.

Roast kid (*French Alps*)

Chop a 1in (2·5cm) long sprig of tansy or thyme, a handful of parsley and two cloves of garlic; crush 1 teaspoonful of dill or caraway seeds; mix the chopped and crushed herbs with 2 tablespoonfuls of wine vinegar, 1 teaspoonful of salt and 1 teacupful of olive oil.

Lay a fresh leg of kid of six to nine months old on a dish. Pour the mixture over it; baste it with this marinade frequently, and turn it daily for 3–4 days, keeping it in a cool airy place.

Pour 1 teacupful of the marinade into a roasting pan. Place the joint in the pan and roast in the oven at 300°F (150°C, gas mark 2) allowing 30–32 minutes per lb (454g).

Ten minutes before the meat is done, melt 2oz (57g) of butter in a saucepan, add a small clove of garlic well chopped; let it simmer 5 minutes; stir in 1 tablespoonful of flour; cook and stir till the flour is golden brown; add ½ pint (0·3 litres) of stock or hot water, and stir vigorously till the sauce thickens; withdraw from the heat, and add 1 tablespoonful wine vinegar, 1 teaspoonful dark honey, 1 tablespoonful concentrated tomato purée – or 2 tablespoonfuls fresh tomato purée – and continue heating till the sauce starts to boil.

Serve this sauce with the roast kid; do not serve the gravy from the roasting dish.

This is kid at its best.

Grilled shoulder of kid (*an old Highland recipe*)

Detach the shank from the blade of a shoulder of kid of three to six months old – so that the joint is shaped like an axe-head. Lay aside the shank. Rub the joint over with butter.

Place on a grilling rack 3–4in (8–10cm) under the grill which should be moderately hot; turn the joint frequently and baste occasionally; test for readiness by pricking the thicker portions with a fine needle; the exudation should be pale pink.

Serve with rowan or redcurrant jelly.

Pekin kid (*adapted from a Chinese recipe: suitable for meat from any age of kid*)

Cut the kid-meat into small pieces, dip in a thin flour, water and salt batter, and deep-fry till golden brown. Lay aside in a warm place to drain.

Melt 2oz (57g) of butter in a large saucepan, add ½lb (227g) of carrots cut in fine strips, and a dozen sticks of celery or seakale beet cut in short lengths; cover and cook slowly, tossing and stirring occasionally. Add water only if necessary. Cook for 20 minutes.

In another saucepan, heat 3 tablespoonfuls of olive oil until smoking hot, stir in 2 tablespoonfuls of flour, and cook till golden brown; add ¾ pint (426ml) of stock, and stir vigorously until the sauce thickens. Withdraw from the heat and stir in 1 teaspoonful of golden syrup, 1 teaspoonful of French mustard, 1 tablespoonful of mango chutney and a little pepper. Pour the sauce over the vegetables, add ¼lb (110g) mushrooms and ½ teacupful of chopped chives, and then the fried meat which was laid aside. Cook for 5–10 minutes. Add 1lb (454g) of finely shredded cabbage, cover and keep barely simmering for 20 minutes more. Serve with rice.

Kid and green peas (*Normandy*)

From a kid about one month old cut small pieces of meat, salt and flour them and brown them in butter in a large saucepan. Add 1 pint (0·6 litres) water per lb (454g) of meat, and stir till blended. Then add 4 spring onions, one 6in (15cm) sprig of parsley, and 1lb of fresh green peas per lb of meat. Simmer gently for 1 hour.

(If tinned or frozen peas are used, they should be added only 15 minutes before serving.)

Kid in cream (*Sweden*)

From a kid of about one month old, cut slices of meat or small joints for serving. Sprinkle all over with salt and pepper and dredge very lightly with flour. Place in a thick saucepan with ½ cup of thick cream per lb (454g) of meat and bone. Cook until the meat is browned, turning frequently, and adding more cream as necessary. Cover, and cook gently until the meat is tender.

Remove the meat into a heated dish, and keep hot. Add sufficient flour to the fat remaining in the pan to make a smooth cream. Cook for 3 minutes, then add equal parts cream and kid stock (or chicken stock), allowing ½ cup of each for every 3 tablespoonfuls of flour added. Stir till the sauce boils. Add chopped parsley, and pour over the meat.

Goat ham (*a traditional Highland recipe*)

The hind leg of a twelve- to eighteen-month old castrate (or sterile goatling) is trimmed to shape and rubbed with the following mixture: 1oz (28g) saltpetre, 4oz (110g) brown sugar, 1lb (454g) preserving salt, 1oz (28g) white pepper, ¼oz (7g) of cloves, 1 grated nutmeg, and ½oz (14g) coriander seeds. Rub the mixture into every crease and crevice in the flesh, and stuff some up the hole in the shank. Lay the ham in a trough, and cover it carefully to exclude dust and flies; baste it with the brine and turn it every day for a fortnight. Then take it out and press it on a draining board for a day.

Remove one end from, and wash out, a treacle barrel or a wooden cask. Make a heap of small birch branches and/or juniper branches and/or oak sawdust in the bottom of the treacle barrel, and bury in the heap a thick bar or lump of red-hot iron; hang the ham from the top of the barrel, and cover to conserve the smoke; punch one or two small holes in the bottom of the barrel to keep the smouldering going. Smoke the ham in this way for as near a fortnight as may be. Then hang it in the kitchen till required. It can be dried by hanging in the kitchen if necessary; but the flavour is much improved by smoking, especially if juniper is used.

The promotion and marketing of goat-meat

'Capra quality goat-meat, a new concept in quality meat eating', said the press information at the 1992 launch of Capra for the British Angora Goat Society. A great deal of work preceded that moment, effort put in because of the importance to fibre and dairy goat producers of being able to sell surplus stock for meat. Capra meat comes exclusively from Angora goats or their crosses, which have been reared extensively or semi-extensively under good welfare conditions, without growth promoters. They must be slaughtered before 24 months and males must be castrates. Carcasses are inspected by an approved grader and are only accepted if classed 2 or 3.

In this way it is hoped to attract the high prices necessary if rearing kids for meat is to be worthwhile financially. High quality meat is also the promotional theme of Goat Meat Producers (Scotland) Ltd, a co-operative whose members, who must live in Scotland, can supply all breeds of goat. The meat was initially marketed by the Fresh Meat Company (FMC) whose cutting plant have pioneered new ways of cutting and presenting 'chevon, the meat of tomorrow' and who have put on promotions of meat and cooked dishes at Scottish hotels. The problem of getting goat-meat into the wholesale market has always been that of providing an even, regular supply, and this problem has been cracked by Goatmeat Producers, who have managed to supply 60 goats a week, 52 weeks a year; on average 81 per cent grade out at classes 2 and 3. This infant meat-market needs to grow and flourish as a support for the industry of goatkeepers and a humane end for unwanted stock.

Legislative control

The massive 'Fresh Meat (Hygiene and Inspection) Regulations 1992' specifically include goat-meat, and they replaced almost all previous meat legislation, to fulfil EC requirements. However, they do not apply to premises which cut and store meat only for sale to the final consumer, which perhaps includes most goatkeepers. This situation involves Registration of Food Premises (*see* Chapter 11), and close control of hygiene and temperatures of transport and storage. Fresh meat for sale must at all times be at or below 8°C and frozen meat at or below −18°C.

The welfare of animals is a subject dear to the hearts, and rightly so, of British legislators, at home and in Brussels. The welfare of animals, including goats, in transit, at markets, and prior to slaughter is all dealt with by new rules. If you take goats to the abattoir it must be in a sound, roadworthy and cleanable box or livestock trailer (not the boot of the car!) which has to comply with detailed, but common-sense, requirements. When you arrive at the slaughter-house, if you do not find that it has closed down because it could not afford to up-grade to EC standards, your goats will be subjected to ante-mortem inspection. After that it is up to the slaughterman to comply with the rules. Goats and kids in livestock markets are a pathetic sight; much better to dispose of unwanted stock in some final and humane manner.

CHAPTER FOURTEEN

Leather and Fleece

Skins of goats and kids

Goatskin provides the raw material for many of the top-quality products of the leather industry. Morocco leather is derived from the long-haired goats of cooler climates. Glacé kid and suède kid shoe leathers are derived from nine- to eighteen-month-old castrates and adult goats of warmer regions. True kidskins, from animals of one to six months old, supply a rather limited market for high-class glove making. A tanner in the north of England produces excellent fine shirt-quality leather.

A kidskin when dried will weigh up to 14oz (397g), an ordinary goatskin 1½–2lb (680–907g) and the skin of an adult male from 3–5lb (1·4–2·25kg). The potential output from Britain's 75,000–100,000 goats would appear to be considerable at present-day leather prices.

Home-produced goatskin is essentially of poor quality; and, in any country, the marketing of raw goatskin is dependent on the existence of a substantial market for goat-meat. Goatskins, like other skins, are no more than by-products of the slaughter-house.

The best skins come from goats on a low standard of nutrition; in practice, they come mainly from goats which are kept primarily for meat production in areas too poor to sustain dairy enterprise. The skin of the well-fed goat is too heavily impregnated with fat to dress satisfactorily; the texture of the skin of housed goats is weakened by abnormal activity of the sweat glands. The best goatskin in Britain is worn by the scattered flocks of feral goats, and this will make up into morocco leather of fair quality.

Nevertheless, goatkeeping is so often a commendable gesture of independence that many goatkeepers will be interested in using this goat product. Though it may prove impossible to produce a first-class article, we can produce, from our own goatskin, sound and serviceable leather goods for our own use at about one-fifth of the cost of buying comparable articles.

The quality of a skin is much influenced by the time of year at which the goat is killed. During the summer the skin of the goat is pinkish and full of small blood vessels, which are conveying into and storing in the skin the nutrients required to produce the denser winter coat. Leather made from such skin is weak and open in texture, and difficult to cure satisfactorily because of the presence of perishable nutrients which give the leather a muddy appearance. When the goat is in full winter coat, the skin is white and the skin nutrients have been transferred to the hair. At this stage, both leather and hair are at their best, and quality goatskins, for use as rugs, can command quite high prices.

Skinning the goat should always be done when the body is still warm; the ligaments attaching the skin to the body of the goat are remarkably strong; if left until cold, it is extremely difficult to remove the skin cleanly. With a sharp knife, make a single clean, light cut from a central point between the two teats to the skin above the breast bone. Do not cut more deeply than is necessary to penetrate the outer skin. From the same point make a cut to the skin above the first joint of the hind legs. Loosen the skin from the belly and thighs, using the fingers and a small wooden paddle, like a flat wooden spoon, but not a knife. (The butcher uses a sharp knife, and does the job more quickly; but for skin and skinner, this is a safer method.) Then carry on the leg cuts to a point just above the hoof. Loosen the skin around the anus and vulva, and cut around these openings. Cut a slit along the centre underside of the tail, and peel the tail skin. Now loosen and peel the skin from the back and flanks, using the paddle to loosen the skin right up to the front of the chest. Continue the first belly cut up to the throat, and work the skin carefully free of the keel of the chest, where the attachment is very close. Cut a slit up to the first joints of the front legs, and strip them in the same way as the hind legs. Cut right round the head just behind the jaw bone and ears, and peel the remaining skin from the neck; it is seldom worth skinning a goat's head.

To remove the hair from a goatskin, make a solution of 4 quarts (4·5 litres) slaked lime and 5 gallons (22·7 litres) of soft water; then stir in 9 pints (5·1 litres) of hardwood ashes. Stir the mixture and soak the skin in it for 3 hours, then hang it on the side of the tub for a few minutes to drain off. Do this four times the first day, three times the next day, and once a day thereafter, until the hair parts on the thickest part; then rinse off in clean water and scrape off the hair.

A method for tanning goatskins with the hair on is as follows: place the skin in a 25 per cent formalin solution for one week. Remove from the solution, and wash in clean water until free of formalin. At this stage, it may be possible to remove surplus tissue from the skin.

Carefully stretch on a suitable-sized board, flesh side out, to dry, using nails to hold the tension (around the outer edge of the skin). When dry, rub in Optimalin ('Lancrolin') oil to soften the skin, and scrape it to remove surplus tissue (this should be done with care). This step should be repeated at intervals (daily) until the skin is clean, soft and pliable. Then wash in household detergent, to remove surplus Optimalin ('Lancrolin'), and restretch to dry, hair side out.

To store skins for curing, prepare a concentrated brine solution by adding salt to boiling water until the water will take up no more; cool, and store the brine solution in airtight containers. Dip the skin to be stored in the brine solution, lay it hair side down on a wooden floor which has been sprinkled with salt; sprinkle the flesh side of the skin with salt; lay subsequent skins, similarly treated, on top of the first, in piles of up to twenty-five skins. When no further skins are likely to be forthcoming for a considerable time, bundle the salted skins by rolling up the pile; tie with string, wrap in thick brown paper to exclude flies and moths, and store for up to four months. Before curing, soak and rinse the skins thoroughly.

To cut a tanned skin to shape, lay it, hair side down, on a wooden table. Mark out with a pencil, and cut with a razor blade, or special leather knife.

Goat hair

Goat hair has a number of commercial uses. The coarse, long hair is, in some countries, used for weaving into tent cloth and for making the basis of carpets and rugs.

As far as the hair of our dairy goats is concerned, short hair is of no interest; but there exists a very small demand for long goat hair, with top prices for white. This is used now to a very limited extent for sporran making.

Angora goats and mohair

Angora goats have been prized for their fine, lustrous fleece since pre-biblical times. The name Angora comes from Turkey, where commercial mohair production was first developed. Today the major producing countries, in order of decreasing weight, are South Africa, USA, Turkey, Argentina, Lesotho, with Australia and New Zealand becom-

Plate 16 Angora goats bred in England. The winning trio at the Bath and West show, 1992. Owner and breeder, Marjorie Jarvis. Photograph: Simon Tupper.

ing established, and the UK producing as yet very little but importing more than any other country to process. The International Mohair Association was established in 1974.

For many years there were no Angora goats in Britain, but following the importation of a small number from New Zealand and Tasmania in 1981, allowed due to their high health status, numbers rapidly increased, aided by embryo transfer. To start with, prices were in four figures and quality was low. Crossing Angora sires with dairy-goat dams produced Cashgora goats, but also a great many low-grade mohair fleeces, liberally mixed with the dreaded 'kemp' the coarse guard-hairs of the dairy-goat's normal coat.

Soon imports from other countries were allowed, notably goats of Texan breeding from Canada. These had heavier, greasier, maybe a little coarser fleeces than the Australasian stock, but free of kemp in the best specimens. Supported by the British Angora Goat Society, breeders have brought about huge improvements in ten years, to produce lovely animals like those shown in Plate 16. Angoras are a little smaller than dairy breeds, and most are white, though there is a demand for coloured mohair for hand spinning. BAGS have a facility for registering coloured animals, and there is a society to promote them. The majority of coloured Angoras are black.

While dairy and cashmere goats have double coats, guard-hairs being produced by primary follicles and down by secondary follicles, in Angoras the primary and the secondary follicles all produce mohair. This grows throughout the year, at about 1in (2·5cm) per month, and reaches a suitable length for shearing twice a year. The production of a fleece uncontaminated with hay – or, a complete disaster, wood-shavings – and shorn before matting commences, requires careful management and is vital to success. Shearing, with hand or electric shears, can be learned by attending the excellent training courses run by the Agricultural Training Board as, indeed, can any other aspect of goatkeeping. An important factor in the value of all the goat fleeces, mohair included, is its fineness. This is measured microscopically as the diameter of the hair in microns. Fibre-processing firms have professional graders who can grade fleeces by eye, this skill taking many years to acquire. It is wise for Angora goatkeepers to practise on samples of known diameter, in order to be able to judge the worth of their own stock. The grades of mohair are as follows:

Superfine kid	23 microns or less
Kid	23–26 microns
Young goat	26–30 microns
Adult	30–34 microns
Strong adult	over 34 microns.

As implied by this grading, as a goat gets older its fleece coarsens. There are uses for all grades, from fine knitwear to conveyor belts, but fine fleece is the most valuable. The prices of all grades, but notably the coarser ones, fluctuate, depending largely on the demands of the fashion trade. Producers in this country can sell their fleeces through British Mohair Marketing, or have them processed into yarn, or spin, dye, weave or knit by hand. Some beautiful mohair mixture fabrics are produced industrially.

Cashgora goats and cashgora

Cashgora goats, as mentioned above, arose from crossing Angora and other breeds. Cashgora was hailed as a new natural fibre, neither mohair nor cashmere, and with a fibre diameter between the two, from 19 to 22 microns. In Australia and New Zealand, cashgora is graded, but the British Cashgora Association have established only one grade, British Cashgora. It is plain from the feel of a handful of mohair or cashmere that diameter is not the only thing that separates them; structurally they are different fibres. Mohair is cool and lustrous, cashmere warm and soft beyond belief. Cashgora is intermediate, warm but slightly lustrous, and makes up most attractively. It is, however, still in the pioneering stage and some do not believe in it at all!

Cashmere goats and cashmere

As explained above, cashmere is the winter under-coat of many breeds of goat. Otherwise known as down or pashm, our own dairy breeds grow some – Anglo-Nubians the least and Toggenburgs the most, but it is too short to process. Breeds which are regarded as cashmere-bearing have a down-fibre length of 1¼–3½in (35–90mm). They have been placed in four groups: Western, such as the Russian breed, the Altai Mountain; Eastern, such as the Chinese Liaoning; feral and northern, such as Scottish and Icelandic; and goats which have been cross-bred with mohair-producing goats to improve the yield, but this is a practice which results in fibre which is not true cashmere. The finest mohair recorded is around 12·5 microns; up to 16·5 microns it is 'hosiery grade' used for knitwear; from 16·5 to 18·5 microns it is 'weaving grade', less valuable and made into cloth. Over 18·5 is rejected by the processors. White down is more highly priced than coloured.

Scotland has been synonymous with the production of luxurious cashmere garments for many years, but these were made entirely from imported fibre. The history of home-grown cashmere begins in the early 1980s when the Hill Farming Research Organization, now the Macaulay Land Use Research Institute (MLURI) used feral goats for studies on complementary grazing studies, alongside sheep. Fine cashmere was discovered on these ferals. In 1986 the Scottish Cashmere Producers' Association was founded, to assist breeding-up of goats, help breeders and market the fleece. At the same time, Cashmere Breeders Ltd was started, a breeding scheme with the élite herd at MLURI, to make the best use of imported genetic material. Goats, semen and embryos were brought in from 1986 to 1988 from Iceland, Tasmania, New Zealand (ferals had also been 'discovered' in Australasia) and Siberia. Funding was successfully raised, and stylish promotions organized. In 1989 the British Cashmere Goat Society was formed as a registration organization. It became affiliated to the British Goat society in 1992.

With all the expertise available, cashmere production in the UK ought to have a bright future. Present problems are the small down-weight per goat – unless the down is too coarse to be acceptable, usually due to Angora influence – and the difficulty of 'adding value'. It is not at all easy to process cashmere as a home industry, because even if it is harvested by combing, rather than shearing, de-hairing is necessary (i.e. removing guard-hairs). Up to now, this is only possible, industrially, on lots over 44lb (20kg), though there is talk of a much smaller machine from America. In addition the production of cashmere garments of professional quality is the result of many, many years of industrial development of a high degree of skill and craftmanship. However, there is no problem in selling cashmere fleece, just as it comes from the goat, for instance through the Scottish Cashmere Producers' Association's fleece pool, which opens every spring. Fleeces are graded and a report sent out, which is an invaluable guide to future breeding plans. Research is under way into the possibility of hormonal synchronization of the shedding of cashmere, which at present can be any time from January to May, in order to make combing more economically feasible.

CHAPTER FIFTEEN

Cropping for Goats

This chapter is intended to help the domestic goatkeeper who wishes to grow a substantial proportion of the food required by two or three milkers. If proper use is to be made of the manual labour involved, and of the restricted cropping space normally available, crop production on such a small scale calls for exceptional accuracy in estimating and applying the quantities of seed and manure needed, and in allocating space to appropriate crops.

The goat cropping ground under such circumstances is liable to be interchangeable with the kitchen garden; so the soil must be treated in a way consistent with high-quality market garden production, and not like ordinary farmland. A high humus content must be maintained, and the fertilizers used must leave no toxic residues. The need to grow field crops by garden methods, in quantities nicely adjusted to the appetite of two or three goats, is not catered for in either agricultural or horticultural textbooks. So a few guidelines are offered below.

A goat giving up to 350 gallons (1,591 litres) a year, which is as good a goat as most of us can hope to own, needs the cropping capacity of rather more than a ½ acre (0·2ha) to supply all her food. A goatling needs about a ⅓ acre (0·15ha); a kid needs about a ¼ acre (0·1ha) in its first year.

In allocating the cropping ground to different crops, and matching the size of the various plots to the goats' needs, it is convenient to use as our unit of reference the 'pole' or 'rod' of 30¼sq yd (25m²). It is easily visualized as the standard 3yd by 10yd (2·7 × 9·1m) vegetable garden bed. One hundred and sixty poles go to the acre (0·4ha); 1 ton (1·02 tonnes) per acre is equivalent to 1 stone (14lb) (6·4kg) per pole (25m²) which eases the translation of field crop recipes. In terms of poles, a good milker needs about 95 (1303m²), a goatling 54 (740m²) and a kid 37 (54m²) – that is to grow all its food, concentrates, hay, the lot.

To calculate the goats' needs in terms of actual crops, winter and summer feeding must be considered separately. In winter, a 180lb (81·6kg) goat in milk will need about 5lb (2·25kg) of hay a day, to maintain health and butter fats, say 2lb (0·9kg) of concentrates, and as much kale and roots as her milk yield justifies. In terms of metabolizable energy (ME), that is 11·8MJ from hay, 9·9MJ from concentrates and 17.4MJ from kale or roots, for a goat giving 1 gallon (4.5 litres) per day. Winter is shorter in the south than in the north, but assuming a 6-month northern winter, the goat is going to need a total of 2,122·8MJ from hay, 1,774·8MJ from concentrates and 3,201·6MJ from kale, etc.

In summer we may allow the same 6 months' concentrate requirement of 1,774·8MJ; the rest of the ration, say 5,324·4MJ, will be wanted in the form of fresh green fodder.

The productivity of land, in terms of ME, varies somewhat with inherent fertility, and the suitability of crop to soil and climate; but, on garden-size plots worked by garden methods, we can assume a high level of fertility; provided we choose crops to suit the district, we can make some useful generalizations. Hay and grain crops yield about 111·36MJ per pole ($25m^2$); green crops, such as grass, cabbage, kale and lucerne, yield about 156·6MJ per pole, and enough protein to balance the ME for milk production. The popular root crops yield about the same amount of ME as the green crops, but with very little protein. Table 19 summarizes the position.

Of course, neither milking goats, kids nor goatlings maintain this conveniently level appetite throughout the year. But, when planning the cropping, it is impossible to forecast food needs with day-to-day accuracy months ahead. The method suggested here maintains a flexible link between supply and demand, and includes a necessary margin for error and for partial crop failures. Many domestic goat-keepers may prefer to buy in their concentrates, or their hay, or both; using this method, it is easy to adjust the cropping and plan accordingly. The productivity of the major goat crops is listed in Table 20.

There is no great difficulty in growing the goats' corn and concentrates in small, garden-scale plots, but special considerations are involved. In the British climate the choice of high-protein concentrate crops is limited, practically speaking, to beans, peas, and linseed. Goats are not always very keen on beans, which have a slightly constipating effect. Peas are tricky in a wet season in a wet district. Linseed, too, prefers a sunny climate, and suffers much damage from wet harvest weather; even in the best of seasons, it is not a heavy cropper, yielding only 62·64MJ ME per pole ($25m^2$) of a valuable, but not very convenient, food. In many districts it may prove advisable to grow the concentrate ration in the form of oats or barley, and to rely on the fresh green foods to provide the protein needed to balance the diet. On the other hand, if a balanced concentrate mixture is bought in, it may be

Table 19 Area requirements of goat cropping ground

	per day	*per 6 months*	*in cropping ground*	
	MJ ME	*MJ ME*	*poles*	*m*²
Needed for a milker:				
(winter)				
from hay	11·8	2,131·5	20	506
from corn	9·9	1,774·8	16	405
from kale, etc.	17·7	3,551·3	20	506
(summer)				
from corn	9·9	1,774·8	16	405
from green crops	29·3	5,327·8	34	860
Needed for a kid:				
(1st summer)				
green crops	11·8	2,131·5	13	329
(winter)				
from hay	7·9	1,419·8	13	329
from corn	3·9	709·9	6	152
from kale, etc.	3·9	709·9	5	126
Needed for a goatling:				
(winter)				
from hay	11·8	2,131·5	20	506
from corn	3·9	709·9	6	152
from kale, etc.	3·9	709·9	5	126
(summer)				
from green crops	19·7	3,551·3	23	582

preferable not to stake too much on the kale crop, which is the key to winter protein; kale too often falls victim to accidents of weather and the persistence of wood pigeons. Fodder beet and potatoes are much more reliable crops for winter feeding. In summer, a succession of mashlum (p. 272) maize and fodder radish can replace lucerne, with a gain in variety of diet and a saving of cropping space – provided the concentrate ration is a bought-in, balanced mixture.

If grain crops are to be grown to provide the concentrate ration, their straw will relieve the demand for hay – if their straw is edible. No allowance is made for this uncertain factor in the table above. By harvesting the cereal crop before it is quite ripe, and feeding it 'in the sheaf', the hay and concentrate ration may be combined – which has

Table 20 Productivity of major goat crops

Crop	*Weight of crop kg per pole (25m²)*	*MJ ME per pole (25m²)*	*Notes*
Hay	13·6–18·1	78·3–109·62	
Oats (grain only)	8·2	87	
(straw only)	12·7	43·5	
Barley (grain only)	9·5	109·62	
(straw only)	9·5	43·5	Seldom eaten
Beans (seed only)	12·7	109·6	Protein-rich
(straw only)	14·1	43·5	
Peas (seed only)	12·7	109·6	Protein-rich
(straw only)	13·6	40	Edible
Linseed	3·2	62·6	Protein-rich[1]
Lucerne (hay)	18·1	87	Protein-rich
(fresh-cut)	95·2	165·3	Protein-rich[2]
Cabbage (drumhead)	108·9	109·6	Protein-rich[3]
(open-headed)	95·2	127	Protein-rich
Kale (marrow-stem or thousand-headed)	95·2	149·6	Protein-rich[3]
Maize (cut when cobs are milky)	81·6	133·9	
Comfrey	136·1	118·3	Protein-rich and very early[4]
Buckwheat	45·3	62·6	Very tasty, for poor ground[5]
Chicory	90·7	78·3	Very early
'Herbal lea' (cut)	81·6	189·6	Cut under 1ft (30cm)
(strip grazed)	60·0	165·3	
Italian ryegrass (cut)	90·7	174	
Nettle hay	15·9	109·6	High-protein
Fodder radish	90·7	127	
Rape	81·6	93·9	[6]
Fodder beet	60·0	149·6	
Potatoes	49·9	158·3	Feed baked or boiled
Mangolds	136·1	142·6	
Swedes	108·9	133·9	
Carrots	59·0	93·9	

[1]*Linseed* is said to cause diarrhoea, constipation, sudden death, goitre and stillbirth.
[2]Fresh-cut *lucerne*, rapidly growing, can cause frothy bloat.
[3]*Kale* and *cabbage* can cause discoloured urine, and *kale* can cause anaemia.
[4]*Comfrey* contains alkaloids which can pass into meat and milk. Feed only small amounts occasionally.
[5]*Buckwheat* can cause photo-sensitization.
[6]*Rape* can cause nervous signs, anaemia and discoloured urine.

advantages for kids and goatlings, and disadvantages for heavy milkers.

Are the goats to be put out to graze during the summer? It will save a lot of labour if they can thus be made to do their own food gathering. But it is very difficult for a few goats to make full use of a small area of ground if they are put out to graze on it. Five goats, put out to graze on 1 acre (0·4ha) of pasture which is the minimum needed to provide their summer keep, concentrates apart, would be faced in July with a patchy paddock of seeded and exhausted grasses, carrying a burden of parasitic worm larvae which would explode into big trouble with each spell of warm, wet weather. This is the most uneconomic way to use pasture. Divide the acre into four and rotate the goats around these four little paddocks, allowing about a week in each, and the results would be better; but much grass would be fouled with droppings and wasted. Strip grazing, with the goats on running tethers (p. 60) or controlled by electric fencing, makes good use of pasture; but the goats may become so fretful with the close restraint, especially when discomfited by hot sun, flies, wind or rain, that they fail to eat their fill, and fail to give their best. With five goats to feed, the balance of advantage may lie with rotational or strip grazing. With two or three goats and similarly restricted acreage, it is often best to cut and carry all they require – in which case, grass, though indispensable at some seasons, is not always and everywhere the ideal crop.

For such crops as are normally grown in the garden, gardening books will provide a wealth of information on the manuring and methods of cultivation; for some of the other crops listed in Table 20 the following notes may be useful.

Oats

These yield better than barley in wet, cold, districts, and provide a more balanced feed than barley anywhere. In wet, late districts a soft-strawed oat, such as Bell or Castleton, is a good proposition, especially if cut on the green side and fed in the sheaf (i.e. unthreshed). In drier districts a grain oat, with a relatively hard, short, straw is generally preferred; Sun II is a good dual-purpose type.
Manuring: A barrowload of compost (not dung) per pole (25m^2), 5lb (2.25kg) hoof-and-horn meal, and 5lb bone meal, (2·25kg) worked into the seed bed.
Sow 12–14oz (340–397g) per pole in March to April, in seed bed with moderate tilth.
Cut, if required green, when the grain exudes milk when pressed; if cutting for grain, cut when the last tinge of green is disappearing from the heads, and leave in stocks or on tripods to dry for a fortnight.

Barley

Manuring: A barrowload of goat dung, 7lb (3·2kg) hoof-and-horn meal, and 4lb (1·8kg) bone meal, worked into the seed bed.
Sow 12–20oz (340–567g) per pole (25m^2), according to the coldness and wetness of the district and the lateness of the sowing, into a fine, deep seed bed from early April onwards.
Cut for hay or green fodder in the flowering stage; cut for grain when dead ripe.

Beans

Use field beans for a seed crop (though runner beans are a productive green crop).
Manuring: 2 barrowloads of goat dung, and 5lb (2·25kg) of bone meal per pole (25m^2) into the drill in late autumn; or spread, for a broadcast drop, in spring.
Sow 1½lb (680g) of seed per pole (25m^2) at 3in (8cm) intervals, in the dunged drills 20–24in (50–60cm) apart, in February.
Cut when the middle pods are just ripe, and dry on tripods. If to be grown mixed with oats, as mashlum, broadcast 14oz (397g) beans per pole into the manured seed bed in February, and 12oz (340g) of a soft-strawed oat 3 weeks later. Cut when the beans are ready. Dry on tripods, set over paper sacks to save the seed.

Peas

Manuring: 2 barrowloads of compost (not dung) with 7lb (3·2kg) bone meal per pole (25m²).
Sow 3oz (85g) of peas with 12oz (340g) of oats, or 4oz (113g) of peas with 10oz (283g) of oats, in a deep, moderate tilth in mid-March.
Cut late August for grain, at the oat-milk stage for hay, or when the first pods are full for use as green fodder. Use tripods for drying.

Linseed (see note to Table 20)

Manuring: a light dressing of compost, or none at all if the previous crop was well manured.
Sow 6–7oz (170–198g) per pole (25m²) in a deep, fine tilth, in April.
Cut when the first bolls ripen, using a sharp blade and leaving a 6in (15cm) stubble; the stem is very tough close to the ground. Linseed must be fed freshly ground, preferable mixed with bruised oats. The crop can be matured in the stock or on tripods; there is always some loss of seed and, as young green linseed is a valuable and tasty feed, it is well to follow linseed with a green forage crop, such as peas and oats.

Lucerne (see note to Table 20)

Manuring: lucerne is a glutton for dung and compost throughout its long life. Use 2 barrowloads of compost per pole (25m²) in the autumn before sowing, 1 barrowload of dung worked into the seed bed in spring with 7lb (3·2kg) of bone meal, then 1 barrowload of compost per pole (25m²) per year.
Sow 2oz (57g) per pole, ½in (13mm) deep, in drills 10in (25cm) apart, allowing 2oz of seed per pole, in April. Soak the seed in bacterial culture, obtainable from the seedsman, if the ground has not recently carried lucerne.
Cut before the flowers fade.

Comfrey (see note to Table 20)

Manuring: comfrey responds well to dressings of dung and compost and hoof-and-horn meal, once it is well established; but in its first year in a fertile soil, no manuring is required; on a hungry patch, a barrowload of compost per pole (25m²) will be sufficient.
Planting: plant the sets 3ft (0·9m) apart, both ways, in ground cleared of

perennial weeds, in April. If the sets are not fresh and vigorous, keep them moist in a box of light compost in a closed frame until new growth is started, before planting them out.
Cut to within 3in (8cm) of the crown of the plant, as soon as flowering stems appear, or at any time when the plant exceeds 18in (45cm) in height. Do not let the plants flower, even if the leaves are not needed immediately for feeding; use unwanted cuts as a mulch or as a compost accelerator.

Chicory

Manuring: chicory is not a greedy feeder, but responds well to dressings of compost applied between the drills in early spring. Work 1 barrowload per pole ($25m^2$) of compost into the seed bed, with 5lb (2·25kg) of hoof-and-horn meal, and 5lb of bone meal.
Sow 2oz (57g) per pole, ½in (13mm) deep, in drills 1ft (30cm) apart, in early May. The plants may be thinned to 9in (23cm) apart, but this is not necessary unless the plot is to be robbed of chicory roots for forcing.
Cut before flowering for green fodder; in the flowering stage for hay.

'Herbal lea' mixtures

For a grazing mixture, buy a small quantity of a standard herbal lea mixture from a recommended supplier (p. 320).
The proportions suggested are as follows (weights are *per acre* [0·40ha]):

8lb (3·6kg) perennial ryegrass
10lb (4·5kg) cocksfoot (or 8lb [3·6kg] timothy on heavy soil)
1lb (454g) rough-stalked meadow grass
3lb (1·4kg) late-flowering Montgomery red clover
2lb (0·9kg) New Zealand mother white clover
1lb (454g) Kent wild white clover
1lb (454g) alsike
2lb (0·9kg) chicory
4lb (1·8kg) burnet (omit on acid soils)
2lb (0·9kg) lucerne (omit on acid soils)
1lb (454g) plantain
16lb (7·3kg) Italian ryegrass

Five oz (142g) per pole ($25m^2$) of this sort of mixture will suffice; or 3½oz (99g) without the Italian ryegrass which is sown as a temporary cover crop. Seven oz (198g) per pole of oats might be substituted for the Italian ryegrass.

For a cutting mixture, something simpler and cheaper will give equally satisfactory results (weights are per pole [25m²)]):

1½oz (42g) cocksfoot (or 1oz [28g] timothy on heavy land) hay strain
½oz (14g) late-flowering Montgomery red clover
½oz (14g) lucerne
½oz (14g) chicory

A little wild seed of plantain and dandelion may be added to this with advantage, provided the neighbouring vegetable gardener has no objection to the appearance of a few dandelion seedlings.
Manuring: 2 barrowloads compost per pole, with 7lb (3·2kg) hoof-and-horn meal, and 7lb ((3·2kg) bone meal, into the seed bed; 1 barrowload of dung to be spread each winter after the lea is established.
Sow in early April or July on to a fine, firm seed bed on a calm, dry day; use the seeding rates given above; roll or tread the seed in, and water it in with a fine rose, lawn sprinkler or hose spray, but do not cover the seed with soil. In April, it may be necessary to offer an alternative supply of bird seed in a convenient container nearby; in July there are ample supplies of bird seed in every field and garden.
Cut the crop when it is 8in to 1ft (20–30cm) high, for green fodder; cut it in the flowering stage for hay.

Kale (see note to Table 20)

Cultivation and manuring as for cabbages. Thousand-headed, 'Canson' and marrow-stem kale are sown in April and crop from September to February; asparagus kale is sown in April and crops from March to May; 'Hungry Gap' kale is sown in July and crops from March to May however hard the winter may be.

Buckwheat (see note to Table 20)

Manuring: none.
Sow 12oz (340g) per pole (25m²), for use as green fodder, in the second week of May. It is very palatable. It may be sown, 6oz (170g) per pole, as a nurse crop for 'herbal lea'.

Nettles

These may be grown in bottomless buckets submerged to their rims in the soil, or in a bed edged with 1ft (30cm) wide corrugated-iron strips (e.g. Nissen-hut link strips), set their width into the soil.
Manuring: dung and coal-fire ashes.
Cut in the flowering stage.

Fodder radish and fodder beet

These require much the same treatment as the related vegetable. Fodder radish is a green fodder catch crop, recently introduced as a big improvement on rape, being more productive, more nutritious and free of the photo-sensitizing effects of rape. It is usefully sown any time from early May to the end of August.

The harvesting of small-scale goat crops can be made easier if some form of drying rack is used to dry such crops as comfrey and chicory and lucerne and grass cut at the green-fodder (under 1ft [30cm] high) stage. A poultry field ark, with a slatted or wire-mesh floor, raised 1ft off the ground to create a good under-draught, performs the function well. A purpose-made ark, using black corrugated-iron sheets to absorb the maximum of sun heat, and providing generous ventilation at the ridge, would probably do the job better. Palatable silage, with less stench than the normal, can be made in plastic bags; fill the bags with the wilted greenstuff, extract the air with the suction pipe of a milking machine, and seal. This is an entirely practical and economic proceeding, but only if you are scrupulously careful in handling the plastic bags, so that they can be re-used. A vacuum cleaner would take a little longer than a milking machine, but if the material in the bags were not too coarse, the vacuum cleaner would be a good tool for silage making.

Calcified seaweed

This is a coral-like substance, which is useful both as a fertilizer and as a food additive. It is carried by the Gulf Stream and found in huge deposits off the coast of Brittany. From there, at a depth of a hundred feet, it is mined and finally arrives in our warehouses and garden shops in the form of a fine white powder, variously known as Seagold, Mermin, etc.

It contains an impressive quantity of calcium and magnesium as well as a wide range of trace elements and, because of its slow action, is safe to apply to the ground, having none of the disadvantages of lime. Farmers using it in all parts of Britain report an improvement in soil structure and fertility as a result as well as an all-round increase in the health and productivity of beef and dairy animals. When one farmer included it in his dairy ration, it saved him from having his milk rejected.

This product is used as a general fertilizer on crops and grass, on farms and gardens. It is included by merchants in some dairy feeds, and a sprinkle of it – about 1 oz (28g) daily – for a dairy goat on concentrate feed may help butterfats. It can also be useful for the vegetable garden and smallholding. Although it is expensive to buy, it is so finely ground that it is economical to use.

CHAPTER SIXTEEN

Goat-farming Systems

To illustrate the operation of principles set forth in previous chapters, here follows a series of descriptions of goat-farming enterprises. These are largely written by the owners themselves, and consequently reflect their own experience and views, which may therefore differ a little from those expressed in the foregoing part of this book. They were all written by dairy-goat owners, but the more extensive methods, such as the last one described, could readily be adapted for fibre-producing goats.

Because these descriptions were written prior to the existence of specific hygiene legislation controlling the production and sale of goats' milk, some procedures described are not now permitted if milk is to be sold for human consumption. Monetary values need to be multiplied by three or four to arrive at today's figures.

Goat farm in West Wales

This is an example of an 80-acre (32·5ha) farm stocked mainly with goats, but with some cattle as a sideline. The farm overlooks Cardigan Bay, only a few miles from the sea but about 600ft (150m) above sea level. The climate is mild and inclined to be showery; heavy rain and snow are rare. The land was described as 'not of the best agricultural type' when purchased seventeen years ago; since then, draining and reseeding have been done with Ministry grants, and rushes are no longer a feature. The clay subsoil is not far from the surface, and water takes some time to drain off many fields, making poaching a problem; but generally grass continues to grow for much of the winter, and good crops of hay and silage are harvested. Ditches and open drains surround the fields, which are mostly of 1½–3 acres (0·6–1·2ha) and surrounded by hedges of beech and blackthorn.

In addition to the goats, the farm has been stocked with a beef herd of cattle-suckler cows rearing their own calves. The farm qualifies for the Hill Cow subsidy, which has greatly helped to improve conditions. It is managed with an adjacent dairy farm by the one owner, although both farms are separately staffed and stocked, with a shared relief. A girl is employed for rearing youngstock and for the goats.

Originally, the main business of the farm was cattle, and it was the goats that were the subsidiary enterprise, so that income from the goats is not easy to establish; but milk and produce have been sold for a number of years, during which it became increasingly apparent that the goats were a worthwhile proposition in their own right. Knowledge had already been acquired of the type of housing, handling numbers, grass seeds to grow, milking arrangements and so on; and the growing demand for goats' milk, cartoned or as yoghurt, encouraged the decision to make the goats the main farming interest. The return of the owner's son to take on the dairy farm and 200 acres (81ha) of the land, leaving the stock farm with 80 or so acres (32·5ha) to carry goats and the remains of the Hereford herd – perhaps ten or twelve animals – has coincided with a need for reappraisal, since the large increase in costs, especially for labour, renders marginal farming unrewarding, unless family owned and managed.

From previous costings, a milking herd of 300 goats is eventually envisaged, producing cartoned milk for a market or supplying yoghurt to a distributing firm. At present, up to 60 animals have been milked, and the numbers will grow within the herd. The costing details of the proposed system have been worked out by the Milk Marketing Board's Low Cost Production service.

To farm this type of land with 300 goats requires a fairly intensive system. Constant daily free range, as at present enjoyed, will only be possible in dry weather after grass cutting, and must be regarded simply as exercise. The empty cattle yards will easily accommodate the milkers, allowing scope for kid yards and male accommodation. The (herbal) grass will be cut and fed in troughs and racks after slight conversion. Rye and other green crops which do well in this part of the country are already grown for feeding in racks. Home-produced hay can, if necessary, be supplemented locally. Straw must be bought in, also beet pulp and dairy rations; oats are purchased in bulk and rolled as required.

Apart from the (cattle) yards, and the smaller feeding space for every animal, there is the milking parlour, which has pipeline equipment, made by Alfa-Laval, for four animals milked simultaneously, with the machine adjusted for goats. Otherwise, equipment and facilities are similar for both herds, as are the standards of cleanliness and tidiness around the parlour, since the dairy must be open to inspection at any time. Individual rations are fed in the milking parlour. General feed and hay, unrationed, are in the troughs.

Kids

As far as possible, the female goats are batch served, so that kiddings occur in groups. The service list is made up in the summer. It is arranged for the first group to kid early in the new year, by utilizing aids available for planning seasons, as with ewes.

Kids stay with their dams for four days; then the dam goes out with the herd, and the kid joins a group of young kids. Each kid is earmarked, disbudded where necessary – under general anaesthetic – and well examined. In the afternoon, the feeder bucket, with teats as for lambs and holding substitute lamb milk, gets each kid sucking. A small group is kept together until kids approach the lambar eagerly and take a teat; then the group will be made up to twelve kids. Hay and water are provided, and a protected feeder hopper supplies the ration – rolled oats and flaked maize, which are frequently replenished. When they are eating well, a coarse protein ration is added. Amounts increase, and the progress of each animal is noted. At two months the kids are still getting substitute milk from the lambar feeder three times a day. When they are judged ready, the lamb-milk substitute is gradually replaced by calf-milk substitute, which continues until the kids are over four months. By then, the quantity of dry food consumed is adequate, taken with hay and water, so the milk feed can be gradually dropped.

It is reckoned that the kid at six months weighs 50–60lb (22·5–27.2kg) and will have cost £25 [1980] to feed. Generally the kids do not go from the open yards. Female kids to be retained are kept together; male kids considered worthy and required for stud purposes are reared in a similar fashion, but continue with substitute milk for as long as they will take it. Two or three are kept together. Unwanted kids are killed at birth. Any to be reared for meat are castrated, run together and weighed frequently.

Before Christmas female kids are examined; and those that qualify in size and maturity will be served by a male kid, these served kids going out for exercise with the adults, but receiving extra rations and being penned together. Before kidding, they will have become used to the parlour by going in and finding rations. The goat must jump on to a raised bench and put her head through a gap to find the bucket. It is found that the newly kidded yearlings enter the parlour and accept the milking machine without trouble, after this practice.

Kidding at fifteen or seventeen months, the animal will have cost £50 to £80 [1980]. Thus, female kids produced in May and June are usually sold, since they will not be mature enough to breed in the first winter, and the chance of their holding to service in March is uncertain.

The system of serving kids is vital to a commercial farm: early kids do better; but, after March, progress seems to take longer. Every way must be sought for cutting costs, not by skimping on feed, but by eliminating wasted food and labour. The animal must be productive as soon as, and as long as, satisfactorily possible. Goats are unlikely to milk economically after ten years of age, so approximately nine years' milking life must pay for rearing and yearly keep, as well as the overheads – there is no carcass value at the end.

Milking Goats

Breed is immaterial; milk potential for long lactation, hardiness and good comformation with well-attached udder are the important factors. The highest yielder is not necessarily the best commercial proposition; an average medium milker gives over 1 gallon (4·5 litres) a day in the first half year, and just under for the rest. This taxes her not at all; and she will give 300 gallons (1,364 litres) for each of her milking years, and a bit more in her prime, without loss of condition. With modern aids for planned seasons, groups of goats could be kidded 'out of season' as a regular system to ensure a constant milk supply from the herd. This may be a future practice; but at present, with care, it is possible to keep a regular supply by running through half the milkers each year.

Males

On this farm the policy is to buy in stud males as kids, with a record of milk and good conformation behind them. The few which are kept from the herd kiddings and the bought ones are reared together, going into adjacent single pens in late August. The kids get well fed throughout their first winter; but dry feed is preferred to continuing the milk substitute, so milk powder is introduced in small amounts in their feed if it should appear advisable. Male kids are used to serve the small females and kids, a few services each.

At the beginning of April, the five or six adult males and the bucklings are put out on a safe paddock with shelter, water, a hay rack and a trough under cover. Since the field is never full of lush grass – for obvious reasons – by the time it grows, hay and feed are refused. They live out all summer, and it is one of the happiest sights on the farm: a gathering of noble-looking gentlemen lying in the sun, with the young lads gathered about them. As soon as precedence is established, they give no trouble at all.

The bucklings come into their winter quarters during August, not, as one might imagine, because they could get hurt by aggressive older males, but because they tend to gang up on their elders as the season approaches: we have had a punch-drunk adult male who most certainly got his condition from two sparring young ones getting his head in between theirs. The older males stay out until September, if the weather is dry.

Housing and Yards

Communal shelter is advantageous, saving bedding, labour and space. The goats lie on raised platforms – a type of kennel/cubicle layout. Straw is used, but very economically; and the 'dung passage' is easily kept clean with a squeegee – a blade of stiff rubber instead of a broom head. Water troughs with ball valves are installed in each yard; they cannot be fouled, and the goats drink plenty from them. It is essential to have adequate room for all together at the hay manger, and that the systems for supplying hay ensure no waste. Minerals in powder form are on offer.

In this arrangement, most of the goat accommodation is in open yards, and health is good. They all have adequate shelter from wind and rain.

Milking Parlour, Dairy for Yoghurt and Cheese-making

It is important to get the goats milked quickly, and the milk cartoned and frozen. This is only possible with a milking machine and bulk refrigerated tank. In fact, we do not believe goats' milk should be sold without this guaranteed way of cooling to below 45°F (7°C) within a very short time. The milk goes directly to the bulk tank from the goat, and chilling commences immediately. As soon as the goats are released, the attendant is free to carton the milk and place it in the freezer. It is believed that milk costs 50p a gallon (4·5 litres) [1980] to produce with the herd at its present size.

Owing to the situation of this farm, markets are some distance away. For this reason, we have a cheese room arranged, and yoghurt is made in a special dairy. Converting milk into yoghurt is worth while; into cheese, less so. The best market is cartoned milk: since the costs of cartoning are minimal, the price can be realistic if it is sold on the yard. A certain amount is collected, and some is delivered with yoghurt and cheese. In summer, local agricultural shows are well worth support, and a great deal of produce is sold, which leads to further orders. By far the best outlet for any goat enterprise must be to sell at the farm to a distributor.

Labour

Labour needs are usually said to be excessive in handling goats: but day-to-day management need not be so. Where most goat owners go wrong is in expecting goats to behave like other stock. Provided the eccentricity of the goat is understood, and the owner thinks one step ahead, problems are few. At a certain time in the afternoon, the herd presents itself at the field gate nearest the farm, waiting to come in; and if, for some untoward reason, their attendant does not arrive on time, they can hardly be blamed for finding a way home on their own.

The routine feeding and milking of possibly 300 goats may produce labour problems. To ensure an acceptable time lapse between milkings, it will be necessary to have one attendant milking from 6 a.m. and doing the morning routine; then another relieving her at noon to do the afternoon routine and evening milking. A third attendant would have to relieve the other two, as well as doing the dairy work. All this is in the interest of the 40-hour week, but gives a labour outlay of well over £100 [1980] a week, plus the little overtime pay that is necessary.

Conclusions

These are difficult to give. The goat has not often been considered a farm animal, especially in Britain; and it is possible that the farm referred to here is the first which will be officially costed. As in all livestock enterprises, the thin margin between profit and loss can be influenced one way or another by as little as the use of half a bale of hay where a quarter is needed, so that a bale and three-quarters a week is wasted.

PAMELA GRISEDALE

Reviser's note, 1992
The Milk Marketing Board's Business Service is now run by their company, Genus (see Appendix 3).

Zero-grazing

While the notion of 'zero-grazing' goats may not appeal to advocates of a natural way of life, this example shows how it may be achieved successfully and humanely. This is no 'factory farm'.

Housing and Yards

To zero-graze successfully, you need a box stall or pen 5ft × 7ft (1·5 × 2·1m) for each goat, and a big exercise yard. The pen should have a hay rack 2ft (0·6m) square, with a lid, so that the goat does not take the food out of the top, and thus waste half of it. There should also be a salt-lick holder with pure salt block, and two rings for buckets – one that falls flat, for the water bucket, and the other firmly fixed, so that the feed bucket can be put in and also taken out for washing. If the feeding bucket is rigid, you can put whole roots in it, thus keeping the goats occupied longer. An idle animal is often an unhappy one, and cannot do her best. Stale bread and kale stalks are eaten well in a fixed bucket; in fact you will find that goats take things back to their fixed bucket after dropping them.

If the sides of the pen are 5ft (1·5m) high, few goats will jump out; but, as they all like to see what is going on, cut a 'window' in the door so that they can put their heads out easily. This stops them from standing on the door and, in time, wrecking the hinges. Board the pens up to 2ft (0·6m) from the floor to cut out draughts, leaving a 2in (5cm) gap between the higher boards so that the goat can see her neighbour; but make sure that the feeding bucket, which may be in the corner, is shut in. Nothing puts a goat off its feed quicker than to have its ears nipped by a neighbour while eating.

For the floor, nothing is better than concrete sloping towards a drain. On this floor you can have wooden slats, which must be made of durable wood, such as elm; they should be in 2in × 1in (5 × 2·5cm), set ½in (1·5cm) apart and nailed on to 4in × 2in (10 × 5cm) crosspieces. These slats are heavy, and get heavier when wet, so it is as well to have them in small sizes – say, four to a pen. Slats are much better for large udders to rest on than cold concrete; also, you don't need so much straw.

It is as well to have at least two yards for the milkers so that you can separate the ones in kid, or for illness. Again, the floor should be concrete, for easy cleaning, slanting to a drain. Concrete should be roughed; otherwise it will be slippery in frosty weather. If you are building from the start, face the goat house door to the south, so that it gets all the sun and the goats are out of the north wind; if possible, build the barn on the east side to keep the goats out of the east wind as well. It is these small details that put extra ounces into the milking bucket.

A branch rack is essential in the yard; it should be made very sturdily to carry big branches that the goats can bark during the winter – and you can burn these as logs the following year. For eight goats, a rack 8ft × 3ft (2·5 × 0·9m) wide, 2ft (0·6m) off the ground and 5ft (1·5m) high is needed. Have the sides made vertical; then the goats must put their heads in to eat and cannot just walk along pushing others out. The yards can be fenced with chain link or 'Weldmesh' or, if you live near a sawmill, with straight elm boards. A shelter from the rain and hot sun is essential: without this, you cannot let them out in all weathers; and they should get out once a day to take in vitamin D through the action of the sun on their skin.

The goat house can be any well-built structure, but it is best to have a fairly high roof to trap warm air; a slate roof with thick insulating boards for a ceiling does well. Provided the pens face a 5ft (1·5m) wide passage, there will be no need for more than the bottom half of a stable door as entrance to the house. This way it will be neither too hot nor too cold; a goat is a tough animal and will stay tough unless coddled. Finally, have a large gutter round barn and goat house to catch all the rainwater; this can be stored in tanks with a pipe through the wall of the barn or food room. A tap with a piece of hose attached is used for filling the boiler to supply hot water. Like humans, goats prefer their drinks hot: not only do they make more milk, but there is no waste of precious food for heating the water in their bodies.

It is better for the male if he can see his females; so build his pen on the end of the goat house. He needs a large house 10ft (3m) square, and a run as long as the other yards. If the kids are housed between his yard and the milkers, he will always have company, and be quieter and easier to handle. Make a feed passage next to his pen so you need never go inside, getting clothes tainted with his smell and contaminating the milk. Have two bucket rings – both fixed, as he will play with them otherwise – with a large salt lick between. The hay rack should be inside his pen, with the back made as a door for filling from the feeding passage. In order to tidy up the yard without his assistance, have a door to his pen, which you can shut from outside; this is arranged by means of a steel bar attached to a sliding door. In the house, instead of littering the whole floor, it is more convenient to have two strong beds of straw-covered slats – one either side, so that he can get out of the wind. These need sides to hold the straw. As long as he is allowed to go in and out at will, the beds will remain dry – which, of course, keeps him much cleaner. A slatted bed in the yard encourages him to lie on that instead of on the wet, cold concrete. He may be found lying out even on frosty nights.

Kids also want a bench outside; and they, too, may lie out in frost. They need a lot of things in their yard for exercising: old water tanks, big rabbit-hutch-sized boxes, and an old tree trunk – if you can get it there – does well. The yard must be big so that they can race about and develop without putting on surplus fat.

Yarding goats is another form of zero-grazing, the only difference being the goats all live together, probably going in and out at will. Pens are arranged for shutting them up during kidding or illness. They have a large communal hayrack but are tied at milking to eat concentrates. High yields need extra rations, so individual feeding is essential.

Feeding

If really high yields are your aim, you must study the goats and their comfort carefully. Let them out for three hours in the morning, while you clean out and refill buckets and racks. Given shelter and branches they will be quite happy for that length of time. To obtain the maximum amount of milk, feed little and often: it is useless to stuff racks twice a day and give them a bucket of cold water.

This is the scheme I use for a day's feed in a zero-grazed herd:

6·45 a.m.	Concentrates during milking; after milking, racks filled with kale in winter, herbal lea in summer.
9·00 a.m.	Milkers let out to large buckets of hot water and branches in racks.
12.00 noon	Milkers let in to hay in racks and roots in buckets (or mangers) in winter, herbal lea in summer.
4·00 p.m.	Racks filled with kale in winter, herbal lea in summer.

6·45 p.m.	Concentrates during milking; followed by hot water buckets, which are left with them for the night; kids are also offered hot water at this time; hot soaked beet pulp is given all year round.
10.00 p.m.	Racks filled with hay in winter, herbal lea in summer.

To make a good milker you must start the day she is born. She wants milk up to six months, rising to 4 pints (2·3 litres) at about a month old, tapering off towards the sixth month. Kids eat sweet, short grass before anything else solid and should have it in their racks at about ten days old. If they are born too early for grass, an old kale stalk with lots of little leaves that are young and tender does just as well. The string with which this is tied up must be well out of reach. Sweet hay should be always with them. Concentrates can be started at about three weeks and, although kids are very wasteful, whatever is fed to milkers must be fed to them, otherwise they will not eat it when adult. Kids must have a salt lick within reach; a pinch of minerals should be added to bottles at a fortnight old, and to concentrates as soon as they are eating well. Soaked beet pulp can be given after they are four months old, but not before, because it is quite indigestible.

All goats, whether male or female, need their feet to be cut every month and should be dusted with louse powder every three months. With zero-grazed herds it is not necessary to dose continually for internal parasites. Once a year is adequate, unless they come in contact with other goats – at shows, say – when more often is advisable. Fed this way, six milkers, two kids and two males can be kept on 1½ acres (0·6ha) of land. The land must be in good heart, however. Of this, 1 acre (0·4ha) should be herbal lea and the ½ acre (0·2ha) used as plough ground: this should include lucerne to be cut green – it is a semi-permanent plant and will stay for four or five years. Roots and kale, winter rye and winter oats should also be grown on the ½ acre. The land must be heavily manured or composted for kale, which is first in rotation; this is followed by roots, succeeded by the rye and oats. The lucerne plot should be heavily manured after the last cut. The herbal lea should be given a thick coating of slurry from any cattle farmer near by; most farmers are delighted to supply this because they invariably have a surplus for which there is no demand. Ten animals do not produce enough manure to fertilize the lea.

In spring, the lea and lucerne are chain-harrowed, then rolled to flatten molehills. Weather permitting, you can take a first cut of rye about the third week in February; this is just when kale is finishing so, with careful management, you can make sure some sort of greenstuff is available all the year round. Rye is followed by oats, then on to the lea, with lucerne last.

You can now stop feeding hay, except when the greenstuff is wet, or at least cut down to one night's feed. The moment rye and oats are finished, the ground is heavily manured and kale sown: we have found 'Canson' kale the best and hardiest. Marrow-stem is liked by the goats but a very heavy frost in November can destroy the lot. Roots are sown in the middle of April, and the kale is best sown in two lots – just in case, in a dry season, thinnings are needed in July or August to keep up milk yields. All kale and roots must be kept well hoed; lucerne also hates weeds and will die back if allowed to get choked. You will have been cutting some part of the lea for greenstuff; this part will grow about three times during the season and give two cuts of hay, providing all the hay needed for ten animals, with a little over. A strip of comfrey is a great standby during a wet season, as it is possible to shake it nearly dry; it yields about five cuts – the same as lucerne.

Breeds and Breeding

To get the best return out of this system, you want big, placid, deep-bodied goats, to give a lot of milk in long lactations. Nothing must be too good for them, or too much trouble. British Saanens or Saanens seem to fit best, since they are content in yards or boxes. Many breeders wait until the kids are about eighteen months old before putting them in kid, but this is a waste of time and money. As long as you decide early, at the kid's birth, in fact, to mate her as a kid and look after her very well, nothing will go wrong. It must be realized that your kid has got to be big, strong and fit; if so, it will be safe to mate her at about seven to nine months old. After mating, feed her extra well, remembering that she has to keep growing as well as feeding two or three kids. Unless well cared for, she will never make a good heavy milker and you will have wasted time and money.

It is, of course, better to breed your own stock, but one has to start somewhere. The best course for the beginner is to make an appointment to see an old-established breeder and take his or her advice. Buying good stock at the start is more expensive, but will never be regretted. Good goats are not cheap: many years of careful selective breeding and exhibiting produce the best. A good goat with a long recorded line behind her could well cost £100 [1980] or more. However, you might be lucky enough to find one with a 'show fault' – say a few coloured hairs on a white goat – and this, put to a really well-bred, milky male could breed useful animals.

Some goats will live and breed for about twelve years, while others have an accident early in life, so buying a young goat is a good proposition. As a rule, what you want at the start is a good milker. Many breeders are overstocked in spring, when most goats kid, and some will sell cheaply to a good permanent home.

By whatever method you finally decide to keep your goats, try to go to a breeder using the same system as the one you intend to adopt. It is no use taking a goat from a zero-grazed herd and putting her out into a lovely field filled with lush grass: she has no idea what to do, so stands at the gate calling to be let in. The idea of eating 'off the floor' will not occur to her and, even if it did, she would not eat enough to give you any decent quantity of milk. It is, however, possible to transfer originally free-range goats to a zero-grazed herd with success.

PHILIPPA AWDRY

Goat Farm near Edinburgh

This is an illustration of the problems encountered in starting a goat farm. I decided to go into commercial goat farming largely as a result of reading an earlier edition of this book in which David Mackenzie argued so prophetically the case for the development of the dairy-goat industry. In fact, we are still a long, long way from reaching the scale of operation that he envisaged, but it is true to say that the prospects for goats are much better now than when he wrote his original words. It is only in the last few years that the interest in goats as a serious commercial proposition has really gathered momentum in this country, apart from the temporary booms created by milk rationing during the World Wars. Now there are several factors which seem to offer longer-term security to the would-be goat farmer: the 'self-sufficiency' movement; the general interest in natural foods and, in particular, the opening of 'whole food' shops; and the gradual recognition by the public of the benefits of goats' milk in the treatment of various complaints. This last, although, in the eyes of the goat's advocates, still pathetically meagre, is certainly the most significant in the long run.

In spite of this, the greatest difficulty the goat farmer encounters is the question of marketing. I only got my operation off the ground at all when I moved from Cumbria to a farm near Edinburgh. The former site offered a brisk, but seasonal, trade with the tourists, but even this depended on the allure of farm-gate sales. Efforts to find retail outlets in the larger Lake District tourist centres met with total failure: attempts to foster an all-year-round local market in the small towns of the West Cumbrian coast brought only a very limited success. At the end of a few years it became obvious that the only way to earn a living from goats was either to mount a massive transport operation or, better, to move closer to the larger centres of population.

So it is only since I came to my present farm, fifteen miles from Edinburgh, that real progress has been made. In many ways, indeed, Edinburgh seemed to be an ideal situation with its high proportion of academics and professional men, plus the fact that there was no large goat herd in the neighbourhood. The latter fact has proved a dubious advantage, as it has meant that the population has no communal experience of goats' milk; no stories of 'wonder cures' for eczema sufferers are part of the local folklore. On the whole we have had to depend on people's natural curiosity, and time: time for the word to get round; time for Mrs A to tell Mrs B what the milk did for her baby; time for Mr X to tell Mr Y about the incomparable flavour of the cheese; time for just a tiny fraction of the population to accustom themselves to the idea that they could buy goats' milk, yoghurt and cheese in their local 'whole food' shop or health store and would actually enjoy it if they did.

There are other reasons why the aspiring goat farmer must allow time for his operation to become viable and why it is a project that should be worked towards over a period. One cannot go out and buy a large herd of goats and establish them overnight. Those who have tried to buy one milking female will have some idea of the impossibility of purchasing a herd of fifty. This can only be done by seizing every opportunity that comes – and those that come will be very few and will have to be sought very assiduously – and by having a carefully thought-out breeding programme. For goat farming to offer any prospect of success, only good – or preferably, very good – stock must be entertained, and the sad fact is that very few goats of either category ever become available for purchase on the open market. The only hope of building up a herd of sufficient standard will be to penetrate the 'old boy' network where most of the transactions involving the right sort of animal are carried out. Here, at least, the novice will have one vital factor in his favour and that is the friendliness and approachability of the vast majority of established breeders.

Three other factors are all-important. One is luck. I bought my first goat unseen from a dealer advertising in a national newspaper. I might have got anything: what I did get was an animal which has given me 300–350 gallons (1,364–1,591 litres) of milk every year since and transmits this characteristic to her daughters. She is unregistered, the only horned female I have ever owned and the only one I ever will own, but she is, apart from that, the ideal commercial goat.

Another thorny problem encountered in developing a suitable herd is the choice of a male. Here the opportunities are far greater because a great number of stud males regularly come on the market. Most of them should never have reached this highly privileged position and are probably doing as much as anything else to hinder the progress of the breed; but there are still some very good males available – and the male, as is well known, is half of the herd.

The last factor is the matter of culling. The breeder for the show ring has long known the importance of this, but it is equally important for the commercial farmer. The profits of goat farming will at best be so slender that passengers cannot be carried. The puny kid and the disappointing milker are luxuries that cannot be afforded. But the commercial farmer has one advantage over his show-ring orientated colleague, who must destroy the animal that is not up to the standard of his herd. The commercial man has an animal which may well be ideal as a family milker, and can be sold as such with a clear conscience.

The fact that the market will be comparatively slow to build up is something of which the goat farmer must necessarily take account because his herd will build up slowly too. We have been working for the last five years to build up a herd of fifty goats, each producing at least 250 gallons (1,136 litres) a year – or, better still, 450–500 gallons (2,045–2,273 litres) in an extended lactation. In another five years, if we are very lucky, we may achieve this target; by then, if not before, the market will need another goat farmer in this area and he will certainly be an ally, not a rival.

Obviously there are many other problems attached to such an infant industry. Finding a suitable farm is a major difficulty. Although there is some flexibility here, it is at least highly desirable that the farm should be capable of feeding the stock, even if the concentrates are to be bought in. This means growing fodder crops to cover most of the year and producing really good hay; this last is so important to the well-being of the whole enterprise that to rely on the fluctuations of the market, and the availability of the right quality, is to invite disaster. Unfortunately, one is still at the mercy of the weather.

Having procured the right farm, with a suitable steading, and about a ½ acre (0·2ha) of good, productive land for each milker, and having budgeted for its purchase or rental, then remember the 'incidentals'. You will need a milking parlour, washing-up room and dairy that will enable you to look the local Environmental Health Officer in the eye; you will need dairy equipment and a milking machine; you will probably need to adapt the housing, build stalls, etc., and to re-fence the land, either to make better use of its resources or to render it goat-proof; you will need a tractor and trailer, at least, and that will leave you at the mercy of contractors; you will need a van; you will need to allow for at least a year or two of running at a loss – in short, you will need the best part of £10,000 [1980], once you have bought your farm or budgeted for its rent. And no bank is going to rush very enthusiastically to your aid.

All this might seem as though this section is addressed solely to the very rich or the blindly optimistic. But this need not be so. Provided there is one substantial wage-earner in the family, ways and means might well be found, provided the final goal is held to be worth the hard work and sacrifice. The most successful goat farm in the world was started by a Californian businessman who was rightly annoyed by the difficulty of obtaining goats' milk for his eczema-afflicted daughter. But if goat farming is really to develop in this country, it will probably find its best recruits from within thc farming industry. Many farms could be profitably developed to include goats as an important sideline; the principals would have the great advantage of having an experience of the basics of general agriculture and, if they were lucky, of dairy procedure too.

But they – and anyone else venturing into this field – will inevitably find that they have much to learn, simply because they are trying to turn into a commercial proposition an animal which has in this country been basically the province of the hobbyist. This is not to belittle the work of the breeders who have made the British goat the finest in the world, but it is to say that many of the methods they have used are not applicable to the commercial farmer. Kid rearing is a good example. The initial problem is of how to feed them; bottle feeding is obviously impracticable with the numbers involved, and we have tried various forms of multiple suckling, none of which seem satisfactory because of the impossibility of knowing just how much milk each individual is taking; we are now trying pail feeding, which at the moment seems to be working well, but we shall have to wait a number of years before we can really judge the long-term results.

Early weaning is also desirable, on grounds of both economy and labour, and is advocated by American writers, as well as by current practice with dairy calves. All the same, it is essential on economic grounds to be able to mate the kids their first season, which we do by retaining only the early kids as herd replacements, and by making sure that they have reached at least 100lb (45·4kg) weight by Christmas so they can be mated as soon as possible in the New Year. The goatling really has no place in the commercial herd, although at the moment we keep a few of the most promising of the late kids through to their second season.

Much of this flies in the face of current 'orthodoxy', although it is common practice in other countries. But, after all, we are dealing with British goats, and hoping to reap the benefits of their excellence; whether we are going to pay in the long run for the short-term advantage, only time and carefully kept records will tell. In this, as in so many matters, the goat farmer has little objective scientific research to guide him: there is only a mass of prejudiced and highly subjective opinion.

There are matters in which the goat farmer faces peculiar difficulties. One is the lack of a stable market price for the stock he is rearing. There is a demand for kid-meat in the high-class butchery trade in this area, but as yet we have not had the opportunity to test its extent and reliability. Ironically, however, the female kids present a greater problem than the males because, in spite of the heavy demand for milking goats, there is very little market for females, however well bred, unless coming from one of the very top show herds. The situation may well arise when potentially excellent stock is either being slaughtered at birth or reared for meat; certainly at the moment we find very little percentage in rearing females who are not wanted as herd replacements. It will be a disastrous situation for the future of goats in this country if good youngstock is being slaughtered, while a broken-down scrub goat attracts hordes of prospective purchasers because she is nominally 'in milk'.

One possible remedy may come when the price paid for a good milking goat more nearly represents her value as a producer, and the cost of rearing her. The goat owner who spends two years lavishing expensive care and attention on his animal and then sells her for £50 or less [in 1977], is doing a grave disservice to goats in this country as well as chalking up a fairly heavy personal loss. In this, as in other matters, the sooner the goat industry is exposed more to harsh economic reality and less to sentiment, the better it will be for goats and the quicker they will achieve their proper role in British agriculture; and in this the self-confessedly commercial orientation of the goat farmer can play a vital role.

There are other matters which are less controversial, but where the goat farmer finds himself without reliable guidelines. Although one or two of the leading animal-feedstuff compounders are producing a concentrate aimed specifically at goats, there is still a great deal to be learnt on this subject. Even more important is a study of the desirable size of the concentrate ration and its effect on milk yield. Goat owners generally have not been orientated towards discovering what is the smallest ration permissible to allow the goat to achieve maximum efficiency; but this is a vital question to the farmer. The housing of the commercial herd also raises the problem of whether the well-being of the animal depends on the initial expense in materials, and continuing expense in labour, of an individual stall. Our present system is to house most of the milkers individually; but we should be the first to admit how inefficient this system is, and as our herd becomes more and more home-reared and therefore, we hope, more compatible, we shall switch to more communal housing.

The British goat is the best in the world. She has been made so by devoted breeders, most of whom have been able to give financial consideration second place. Probably much of what is written above will strike them as quite unworthy of the animal that they know. But the challenge facing the commercial farmer is to adapt this somewhat pampered animal into the useful economic unit she undoubtedly is without losing her good characteristics. The commercial goat will probably never achieve the startling records of some of the great yielders of the past, but a consistent rate of production at a much lower level than this can be profitable within the qualifications already specified. There does seem scope for the development of the dairy goat industry, which can make useful, if minor, contributions to the health and economy of the country, provide the farmer and his family with a modest living, and install the goat in her rightful place in twentieth-century Britain – as a useful producer of a beneficial food.

ROBERT HASLAM

Marginal farm in Wigtownshire

This example shows how a few goats can be usefully integrated into the workings of a small mixed farm. The farm, with an area of 60 acres (24ha), is in a high rainfall area, with very poor soil – sand on rock with an area of peaty bog, plus large patches of gorse (whins in Scotland) full of rabbit burrows. The only trees are hawthorns and the situation is extremely exposed to the prevailing south-west wind. Dry-stone walls surrounding the farm and dividing fields afford some shelter to sheep and out-wintered Galloway cattle. The original excellent solid stone farm steading included a large byre used for storing hay, and a goat house with four boxes roughly 5ft 6in (1·7m) square and one 5ft 6in by 10ft (1·7 by 3m) for kids. In one corner were the milking bench and scales and, leading out of the goat house, the dairy area containing a large sink with hot water from an electric geyser, and a worktop to take the separator and other equipment. Half of a two-stall stable adjoining became the food store, the second stall becoming a general farm emergency room. The building was extended with passages and small yards to hold two large farrowing pigs, while the male goat house, with a large exercise yard, was established within the old stock yard. From here the billy sees the comings and goings of people and animals; he is kept in with a 7ft (2·1m) wire fence with posts concreted in to control both goat and pig.

A 12ft × 10ft (3·6 × 3m) shed with half door and skylight was built of rough forestry slabs, the outer ones set vertically, the inner horizontally, with straw packed tightly between. Though this was originally intended for sows and litters, it became the kid house. Outcrops of rock make a splendid playground for baby kids; an area is fenced off with 1½in (3·8cm) mesh netting with one electric wire above. Kids go to this house and run after leaving their dams at four days and stay here for two weeks before rejoining the herd. After this interval, they are immediately recognized by their dams, but no attempt is made to suck. Kids return to their own quarters when the goats are housed. In reasonable weather they are only shut in last thing at night and can be seen playing the wildest games till late evening. It is autumn before they go to live in the goat house.

Ducks (Khaki Campbells) have been introduced and a duck house has been built near the bog: ducks are believed to eat liver fluke. The loss of a ewe from fluke prompted the idea, but whether or not the ducks have been the reason for no further case in either sheep or goats one can only speculate.

The menace from foxes, prowling round early in the morning waiting to snatch a lamb while its twin was being born, was overcome by keeping the ewes in a field near the house until the lambs were a week old.

After a soil analysis, slag and ground limestone were applied and each year 2 or 3 acres (0·8 or 1·2ha) is sown to oats, undersown with 'Clifton Park' lea. The ground is cleared by sows – folded by an electric fence – which energetically tear out bracken and other weeds, fertilizing the ground as they go. We have had great crops by using this method, grazing the first year after harvest and subsequently cutting for hay.

The Proctor tripod system, far superior to the usual fixed tripod, is used for harvesting both hay and oats. Hay is built on wires and air vents; the latter are then pulled out to leave an almost vertical hut-shaped rack with the hay hanging just clear of the ground. Air circulates in the wide space inside and rain is easily shed by the steep sides. Since

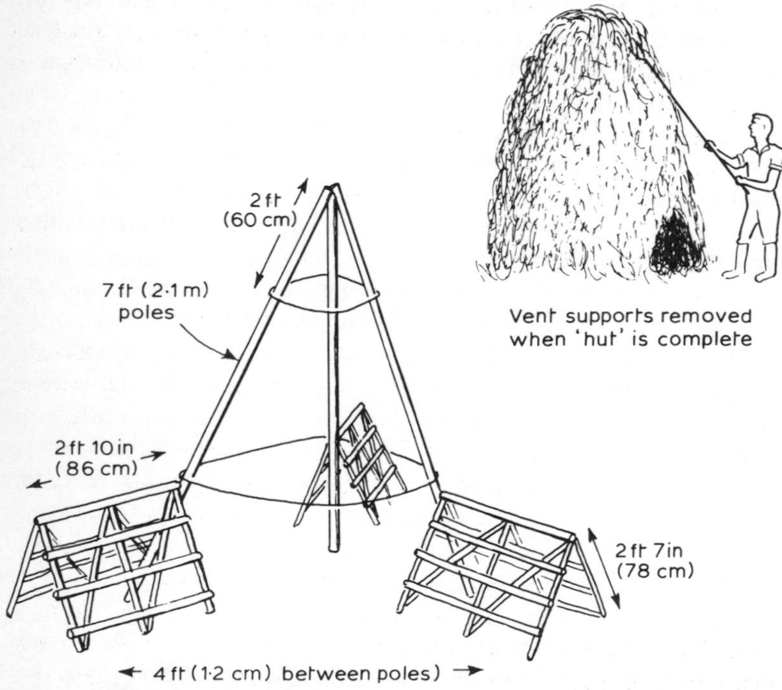

Fig. 21 The Proctor tripod system.

you cannot rely on perfect weather for the baler to pick up hay in mint condition three days after cutting, tripods are the answer – and there is no comparison between the Proctor tripod and any other system for obtaining well-got hay, which is so important to the dairy goat. Small areas can be cut as convenient for the requirements of the small-scale goatkeeper and quickly tripoded. The deep-rooted leas defy drought and the hay ensures a supply of minerals and trace elements to grazing stock. The goats have always milked very well on this diet.

Electric fencing is used to protect the hay and corn ricks while the goats make the most of the aftermath. June hay-making allows the regrowth of fresh high-protein grass to offset the seasonal drop in protein and in milk; and a fresh boost comes after harvest with September weather and aftermath grazing producing milk yields as high as at any time. The absence of flies and hot sun at this season are much appreciated by grazing stock and they seem to get their heads down and instinctively stoke up against the lean months to follow.

Potatoes are grown; the undersized 'chats' are fed to sows together with separated milk and a small dash of meal. Cooked potatoes are also fed to goats at mid-day in winter. Other feed crops are kale ('Canson' is recommended as the best), carrots, fodder beet, and cabbages for winter keep.

Goats are the first on to any new lea, followed by sheep and then by Galloway cows and calves. This system breaks the worm cycle and, once it is established that droppings tests have produced negative worm counts every time, the goats have a linseed-and-turpentine drench in the spring and garlic tablets at the rate of about two twice a week, raised to a daily dose for two or three weeks every spring and summer. Ewes and lambs have a worm drench in spring.

Stocking on the farm is at the rate of three milking goats, with probably two followers; ewes started at ten and worked up to forty-five on the improved grazings, and there are five or six Galloway cows with calves sold off at the autumn calf sales. A bull is hired for six weeks to get them in calf. There are four breeding sows – pedigree large Whites and two litters annually – and a dozen laying hens (RIR x LS). Apart from the ducks, the picture is completed by geese: two and a gander (but they have nearly always had nests robbed by foxes).

Numerous adders sun themselves on rocks: goats have never been bitten, dogs escaped by some miracle, and the only (non-fatal) casualty so far has been one of the farm cats. Permanganate of potash, always at the ready, is effective if applied to a bite within a short time. On another farm a dog, bitten in the throat, was treated with dry crystals followed by a solution and it suffered no ill effects. Another, treated after a time lag of at least twenty or thirty minutes, with solution only, was all right and there was no swelling or subsequent discomfort. Unless provoked, an adder slips away, which is probably the reason why stock are seldom bitten.

In conclusion, the goats, thanks to clean ground and the advantage of deep-rooted leas on ground free of chemicals, herbicides and pesticides, have enjoyed complete freedom from disease without resort to drugs and vaccines. Concentrate feeding, even for goats giving 1¾ gallons (7·9 litres) plus, has never exceeded 2½lb (1·1kg) daily, with no high-protein cake. Seaweed meal and iodized salt licks were the only mineral supplements.

Goat's milk is invaluable on a farm to other stock: it puts a bloom on weaner pigs, which enhances their value, while lambs, calves and foals thrive on the milk, which is suitable for all stock and is far safer and more digestible than cows' milk. The household benefits from all the derivatives: butter is easily made in a bowl with an egg whisk, a most palatable cottage cheese involves the minimum of labour, and yoghurt needs no other equipment than a vacuum flask (see p. 245). The inclusion of goats into our scheme has worked entirely to our advantage – once it was accepted that twice a day for 365 days of the year, the goats must be milked and tended.

JEAN LAING

Environmental issues

Public opinion today is 'green', i.e. there is great pressure that farming systems must conserve, rather than destroy, our environment. It has become a major political issue and, in the current mania for legislation, laws have been passed which farmers ignore at their peril. Goat farming enjoys a 'green' image, but nevertheless care must be taken to avoid polluting the surroundings.

In 1990, the government published a white paper, 'This Common Inheritance' which promised codes of good agricultural practice. The first, for the Protection of Water, emerged in 1991, and has a statutory basis, i.e. contravention of the Code can be taken into account in any legal action, though it is not itself an offence. The Code 'gives farmers practical advice on how to avoid water pollution from farm wastes, fertilizers, nitrate, fuel oil, sheep dip, carcass disposal and pesticides'. It is particularly damaging to river life if water is polluted with substances with a high Biochemical Oxygen Demand (BOD). Milk has a massive BOD, many times higher than other effluents, and surplus milk must never be poured into a public drainage system. It should be diluted and poured on to land which is nowhere near any drainage or waterway, not flooded or frozen, etc.

The second Code is for the Protection of Air. It advises on 'how to minimize air pollution from, for example, odours and dark smoke'. The Environmental Protection Act 1990 gives local authorities a duty of inspecting their areas for 'Statutory Nuisances' one of which is 'any animal kept in such a place or manner as to be prejudicial to health or a nuisance'.

The third Code, for the Protection of Soil, published during 1993, is a practical guide to help farmers and growers avoid long-term damage to soils. It has information on fertility degradation, contamination and restoration of soils.

CHAPTER SEVENTEEN

The Export and Import Trade

Legislative Control

The basis of the European Community is that trade should take place freely within it, and that trade with countries outside the Community (third countries) must not be on more favourable terms than within the Community.

This has generated mountains of paper, covered with rules designed to create confidence in purchasers that, say, a kilo of carrots will be identical be it in Lisbon or London, and that vendors will be on the much discussed 'level playing field', that is that competition between rival businesses shall be fair. One of the biggest problems facing free trade in animals and their products is to avoid an equally free trade in animal diseases, as there are several illnesses restricted to certain of the Member States only, and everyone would like to keep it that way.

So health conditions have been laid down (Directive 91/68) which sheep and goats must satisfy before being sold from one into another European country, and which they must satisfy (Directive 91/69) before they can be imported into the Community from a third country.

In *Directive 91/68*, distinction is made between animals for slaughter, for fattening and for breeding; the health certificates are different in each case. All goats must be identifiable and must, on inspection within forty-eight hours preceding loading, show no clinical signs of disease. They must also comply with conditions referring to brucellosis, rabies, anthrax and foot-and-mouth disease. Animals which are the subject of health restrictions in their own countries may not be traded and if they have come from outside the EEC they must comply with certain conditions. Goats for breeding or fattening must additionally, in order to go into holdings free from brucellosis, comply with conditions laid down which involve a testing regime, etc.

Goats for breeding, which after all is the main purpose for trading in them, have to comply with conditions which in some respects are felt in the UK to be inadequate; for the precautions against the importation of *Mycoplasma aglactia*, paratuberculosis, caseous lymphadenitis, pulmonary adenomatosis and CAE hang largely on the appearance of clinical signs and on an owner's declaration of freedom from these. It is therefore very important for individual goatkeepers to exercise the option no longer open to government, and that is to insist on blood tests or whatever means are available to ensure that any stock they import is truly as healthy as it looks. It is only where a country can claim national freedom from a disease (e.g. *Mycoplasma agalactia*) or point to an official eradication scheme (e.g. the MAFF sheep/goat health scheme for CAE accreditation) that the EC may allow relevant import restrictions to safeguard national health status.

There is one safeguard however, which involves more than blood-testing. Scrapie is Officially Notifiable (since 1 January 1993) though initially at least, this involves no attempt at eradication. Furthermore, breeding goats being traded have to come from herds officially free from scrapie for at least two years. Consequently MAFF opened Category 4 membership of the Sheep/Goat Health Scheme to enable goatkeepers to fulfil this requirement. The cost of membership, and the expensive laboratory examination of the brains of culled goats – the only way to diagnose scrapie – present a very real barrier to 'free trade' in breeding goats between member states, and is the despair of Irish goatkeepers, who have to obtain full health certification to take a goat, even briefly, from Ulster to Eire or vice versa. Any relaxation of standards will spread disease, however; already we have acquired CAE and caseous lymphadenitis.

It is crucial both for the ability to safeguard imports and to use raw milk for human consumption, that we can prove, in Brussels, our national freedom from infection with *Brucella melitensis* (the cause of human Malta fever) and MAFF are testing the requisite number of blood samples to maintain this proof. It is hoped that it will be possible to establish a satisfactory test for caseous lymphadenitis, and include it as an option in the Health Scheme, to control this recent introduction before it spreads out of control.

Directive 91/69 ensures that imports of goats from third countries – countries outside the European Community with which EC countries are trading – will be permitted only from those countries or parts thereof approved by the Community, and each consignment will be accompanied by an official health certificate, the conditions of which will be set by the Community in relation to the country or area concerned.

Products of animal origin, goat semen and ova, have their own health conditions which must be achieved prior to trade between Member States or importation from third countries.

Once standards have been reached, certification supplied and trade is actually in progress, a different set of directives comes into play, the 'Veterinary Checks' Directives.

Directive 90/425 replaces frontier checks, no longer permitted in the Single Market, by checks at points of origin and spot checks at destinations, for live animals being traded within Europe. The central core of the system is the health certificate travelling with the animal, issued by the vendor's inspecting veterinarian. The means of transport has to be suitable and hygienic, though welfare aspects are the subject of a separate directive. A computer system, 'ANIMO', links all member state veterinary units, and information on animal movements between states or into the Community from outside are to be fed into this.

Importers must give twenty-four hours' notice to their MAFF Divisional Veterinary Officer, and when the consignment arrives they must check identifying marks and health certificates and isolate the new goats until told otherwise by the DVO. Certificates must be kept for at least twelve months, and adequate access must be given to the DVO for inspection – the same goes for genetic material. Intra-Community importers and exporters have to register with MAFF. Quarantine for foot-and-mouth disease ended for animals imported from Europe on 5 August 1992.

There are safeguards built in to the directive to deal with any outbreak of serious disease in a member state. The directive also requires that all goats be traceable to their holding of origin, and this extends to movements within the member state. Trade within Europe in products of animal origin, such as semen and embryos, is controlled by Directive 89/662; imports of live animals from third countries by Directive 91/496 and of animal products by 90/675. All these Directives will be implemented by legislation in the UK, making it an offence not to comply with any of the provisions laid down.

Registration of all UK goatkeepers. It was mentioned above that Directive 90/425 requires that *all* goats must be traceable to their holding of origin. This is being followed through in a Regulation Concerning the Identification and Registration of Animals. This requires *all* goatkeepers, even those with one kept as a pet, to supply their names and addresses to their 'competent authority' (e.g. MAFF in England). Once registered in this way, they will be issued with a

'Holding Number', applicable to cattle, sheep, goats and pigs. This has to be marked, with a tattoo or ear tag, on any goat which leaves the holding. Goatkeepers must also record all births and deaths, as well as movements, and hold such records in readiness for inspection. Animals in transit have to be accompanied by a movement document, which has to give certain information, and from 1 January 1995 this applies to journeys within the country, not just for export. Up-to-date records of female goats must be kept, plus an annual stock-take, on a certain day, of all goats.

The quality of livestock

It is evident from the previous section that importing goats is neither easy nor cheap. There must be compelling reasons for us to wish to import, and for those overseas to wish us to export. Almost invariably, that reason is to acquire a genetic make-up which results in greater productivity, be it of milk or fleece or meat.

The selection of stock by would-be importers has been in the past, and no doubt will continue so in the future, by personal contacts, visits to overseas breeders and to agricultural shows. The goat organizations assist by making contacts, issuing export pedigrees, covering the registration of imported stock and their progeny in their regulations, and so on.

A more structured approach is provided by the Ministry of Agriculture's External Relations and Trade Promotion Division, which includes goat information in its information stands at overseas trade fairs, and passes enquiries back to the relevant goat organizations. ERTP also gives advice, and maintains a database of would-be exporters, which is used to provide contact lists of UK businesses for overseas buyers.

The EC saw, in the selection of breeding stock for trade between Member States, yet another opportunity for legislation, and produced Directive 89/361, concerning Pure-Bred Breeding Sheep and Goats. This defines 'pure-bred' and requires that an animal deemed to be pure-bred by an approved breeders' organization in the exporting country, must be eligible for registration in the herd book of an approved breeders' organization in the importing country. The semen, ova and embryos from pure-bred stock are treated in the same manner. The Directive also required that, before 1 January 1991, the criteria for approved breeders' organizations which maintain herd books should be laid down, also the criteria for registration of goats in the herd books, also approved methods for monitoring performance and assessing the genetic value of pure-bred goats, and the criteria for the approval of a breeding animal for the purpose of using its semen, ova or embryos. This was all accomplished by a series of Commission Decisions, 90/254 to 258 inclusive, the last being the form which the various zootechnical certificates must take. Subsequently the Zootechnical Standards Regulations 1992 made it an offence to fail to provide the relevant zootechnical certification when selling EC Approved pure-bred breeding livestock, or genetic material from them.

It was always stressed by MAFF that seeking approval is not mandatory, and that it is perfectly possible to continue to trade without approval. Largely because the definition of 'pure-bred' laid down by the EC is less stringent than that decreed for dairy goats by the British Goat Society, that organization decided not to apply for approval. The British Angora Goat Society, however, did become an EC-Approved breeders' organization, with all that that implies.

The International Goat Association

In the words of its publicity leaflet, 'The IGA is a non-profit organization founded in 1982 with the following objectives:

- To foster the use of goats to provide for the needs of humankind
- To encourage research with and development of goats to increase their productivity and usefulness throughout the world
- To perpetuate the International Conference on Goats
- To sponsor the *Journal of Small Ruminant Research.*

The IGA board of directors includes famous names in goat research from all round the globe. Members receive the *Journal of Small Ruminant Research*, which chronicles the endeavours of scientists around the world to find ways of managing goats better, more cheaply, more productively. The first International Conference on Goats was held in London in 1964, others have followed in Tours, Tucson, Brasilia and New Delhi. The goatkeeping world is shrinking and the potentialities of the different breeds are becoming better known. It is therefore a great tribute to British breeders that their stock is still in demand around the world.

What is needed at both ends of an export business is knowledge. The breeder for export requires to know how his stock perform under foreign conditions, in what way they excel and fail; the importer requires to know more about the stock he buys, especially about those inherent characteristics which are of small importance in this country but great importance in others – fertility, growth rate of kids, etc. Neither of these requirements can be met without controlled recording at both ends and an interchange of views. Tight organization of breeding at both ends of an export trade is an essential to success.

It is more than probable that the accentuation of qualities required by overseas goat farmers may make the British goat and the British Goat Society more interesting to farmers in this country too.

The export trade is variable. During 1976, ninety-two goats were exported through the British Goat Society to Nigeria, Trinidad,

Plate 17 R124 Berkham Butterlady, Q*7 BR. CH. Miss E. Rochford's very elegant Anglo-Nubian. (Photograph: Diane Pearce)

Jamaica, Eire, Northern Ireland and Canada. These were mostly Anglo-Nubians, Saanen and British Saanen, with fewer British Alpine, Toggenburg and British Toggenburg. In 1977, enquiries came again from Eire (British Alpine) and from Nigeria (British Saanen) and additionally from St Vincent (Anglo-Nubian), India (Saanen and British Saanen), Brazil (Anglo-Nubian) and Thailand (British Alpine and British Saanen). Since its political upheaval, Portugal has shown renewed interest. Previous years have seen goats going to Bermuda, France, St Helena, Barbados and Dominica. Private sales include a batch of sixty to Sudan (Anglo-Nubian, Saanen and British Saanen) in 1976; previously, goats of various breeds have been sold privately to France, South Africa, Kenya, Japan and Malta; and lately Anglo-Nubians to Oman, and others to Yemen.

In 1979 about 250 goats were exported from Great Britain, but the main consideration for 1980 was felt to be price. West Germany was one of the main exporters of goats and British exporters for a time found that the high cost of British exports made them less than competitive, despite the fact that goats all over the world have become immensely popular, particularly in developing countries, which have come to realize the value of the goat over the cow.

In the early 1990s, the number of exports was depressed, partly due to some ultra-cautious governments banning imports of goats from the UK because of the cattle disease, bovine spongieform encephalopathy (BSE), which is somewhat ironic when one considers the very high health status of our island population of goats. However, more than forty animals did leave our shores during 1991, groups going to Uganda under the auspices of the 'send a cow' project, and to Kuwait to replace some of those slaughtered by invading Iraq military.

Goat breeders have been urged to make the most of export opportunities by having figures for lactation yields of milk – high protein is sought after – and information leaflets on their breeds at the ready to welcome overseas visitors at shows.

CHAPTER EIGHTEEN

Harness Goats

The Harness Goat Society, affiliated to the British Goat Society, was founded in 1987 to protect and promote the working goat, rarely seen since before World War II, when many seaside promenades featured goat-cart rides to small children.

Society members are in great demand to provide displays at agricultural shows, where turn-outs like that shown in Plate 18, as well as miniature horse-drawn tradesmen's vehicles, delight the crowds. Goats of all breeds and both sexes can be worked, but castrated males are recommended, both for driving and as pack animals.

The Society's booklet, 'Training your Goat' describes the stages of breaking the goat to harness, which have similarities to breaking in a horse.

Long-reining

The goat wears a bridle, to which are attached two reins long enough to be held by the trainer walking behind the goat. Many short lessons are much better than fewer longer ones, to avoid boring the young goat.

Sets of harness can be purchased from smallholder equipment suppliers. The bridle does not need anything but the head-piece and nose-band, and a bit if you need one. The nose-band must allow breathing, of course, and, unlike the horse, room for the goat to open its mouth and pant. Just as with a horse, the goat is taught during long-reining to obey the voice, to make it go forward and stop, while rein pressure is used to turn left or right. Initially, a leader to walk beside the animal is a great help. The driver must praise the goat when it acts correctly. As goats return home much more readily than they leave it, it helps to lead the goat away and then drive it home, to begin with.

It is not natural for a goat to go in front of a being that dominates it, so patience in training is essential. If training can start when the goat is a kid, take it for walks on a lead and try right from the start to get it to walk slightly in front of you. One person should do the training so that one voice can be learned. Say the goat's name, then 'walk on' or 'whoa'.

Plate 18 Harness goat 'Tilly' in action. Owned and trained by Elspeth Pratt. (Photograph: *Sunday Mercury*).

Always stop a session when the goat has just done something right.

The reins are held in the left hand, with the thumb uppermost, the order being: thumb, near-side rein, two fingers, off-side rein. The spare ends of the reins hang down from your palm. If it is necessary to exert more pressure on either left or right rein than can be done with movements of the left hand, the fingers of the right hand may be used. The right hand also carries the driving whip, comfortably balanced between thumb and palm. The whip is used to reinforce the voice. The reins are never flapped on the goat's back; this is for people driving tradesmen's ponies in films! An even feel must be kept on both reins; the goat has a natural tendency to turn to face the driver!

The saddle

When the goat is long-reining well, the saddle, a pad which fits behind the withers, can be put in place and the girth fastened. The saddle carries the tugs through which the shafts are held, and a belly band, looser than the girth; also the backstrap, which runs along the backbone to the goat's hips, where it supports the loin strap, though this and the backstrap are not needed unless full breeching is used. The saddle also bears two rings or terrets, as does the neckstrap in front of it – the reins should be passed through the neckstrap and saddle terrets on the appropriate side, though the neck-strap, which supports the breast collar and traces, may not be fitted till later in training.

The pull on the reins will feel different to the goat with the reins now passing through the saddle terrets, and the long-reining training will have to be repeated. There are 'handy goat' classes at shows, long-reining through an obstacle course, either with or without the vehicle, so it is well worth spending a lot of time working at this.

Finally, the great moment comes: the proficient goat is kitted out with breast collar at the right height, tugs level and traces the right length and the vehicle ready.

Putting to the vehicle

Bring the vehicle carefully up with the shafts either side of the goat, in the tugs. The traces go between the saddle and the belly band each side and hook on to the trace hooks on the vehicle. The breeching straps are fastened to the shafts, and the belly band, tugs, traces and breeching correctly adjusted. Finally, the reins are put in place and off you go (Fig. 22).

The backstrap, loin strap and full breeching can be replaced by false breeching, a strap which goes across from shaft to shaft at the correct distance just behind the goat.

Competition classes

The following are recognized by the Harness Goat Society:

1. Pack Goat/Travois (pole sledge)
2. Private turn-out, led
3. Private turn-out, driven
4. Trade/commercial turn-out, led
5. Trade/commercial turn-out, driven
6. Driven test, long-reined
7. Driven test, sat in vehicle

8 Handy goat, set obstacle course, goats to be long-reined
9 Accuracy test, long-reined through set course of cones
10 Accuracy test, driven through set course of cones.

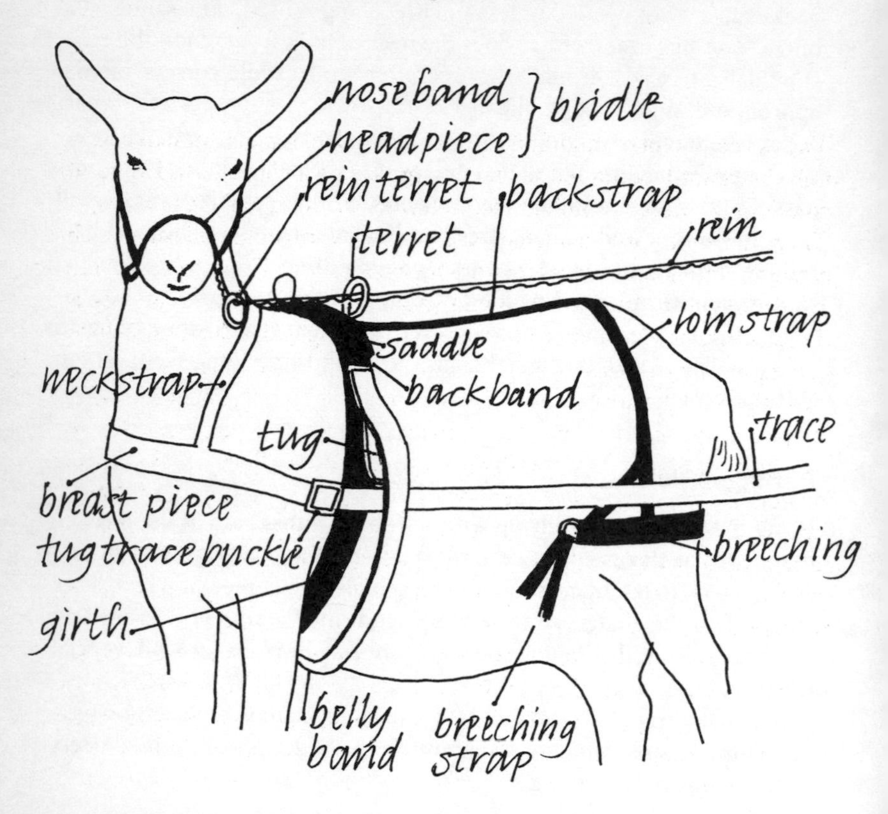

Fig. 22 Diagram of main parts of goat harness.

CHAPTER NINETEEN

Cows'-milk Allergy and Goats' Milk

The development of modified dried cows' milk to replace human breast milk in infant feeding led to the emergence of a clinical syndrome of cows' milk protein intolerance, referred to commonly as 'allergy'. There are other proteins which are involved with an intolerance, such as gluten intolerance which causes coeliac disease. These intolerances are presumed to have an immunological basis. Milk protein allergy is therefore quite different from lactose (milk sugar) intolerance, which occurs in many adults.

Antigens, substances which the body 'regards' as potentially harmful, stimulate the formation of antibodies. Antibodies are specific to particular antigens and bind on to them, neutralizing their harm. Antibodies are proteins of the globulin type, known as immunoglobulins, Ig for short, of which there are five main types, each known by letters, such as IgA and IgE. Protein molecules are large, and at about the time of birth or very soon after, the gut wall 'closes' so that intact protein molecules cannot be absorbed from the food into the bloodstream in any but tiny amounts. (This is why colostrum must be fed so soon after birth, while the mother's Ig molecules can be absorbed by the kid to give passive immunity). The tiny amounts of food protein that are absorbed, however, cause antibody formation. Normally IgA is produced which does not cause problems, but in allergic babies, IgE is formed, and itching, wheezing, etc., occur as a result.

At present 7·5 per cent of babies are now regarded as having cows'-milk protein intolerance. Some become desensitized with age, but in other cases the allergy persists into adult life. Even breast-fed babies can suffer from cows'-milk allergy, since some of the cows'-milk protein the mother drinks will be secreted in her breast milk. The clinical manifestations of cows'-milk protein intolerance are many and various. They can be gastro-intestinal, such as vomiting, diarrhoea, colic, etc.; respiratory – nasal catarrh, chronic cough, bronchitis, asthma, sinusitis, etc.; dermatological – eczema, various rashes and possibly psoriasis; neurological – behaviour disorders, sleep problems, migraines, etc., or others which may include cot death.

Diagnosis of food allergies is difficult. The usual way, other than by skin tests, is to 'challenge' the patient by withholding the suspect substance from the diet and then give a small amount and see what happens. This could be too dangerous in cases where cot death might result.

Once cows'-milk allergy has been established, a diet free of it must be established, and goats' milk has an excellent track record in this respect, though it is possible to be allergic to the protein of both cows' and goats' milk, in which case soya milk will have to be used.

Babies under six months cannot be fed on either cows' or goats' milk without proper modifications as their kidneys are too immature to cope with the mineral content, and brain damage can result. It is always wise to consult a doctor, but the advice of one paediatrician is that first, goats' milk for babies must always be boiled for two minutes and then cooled, immediately prior to use. Milk for babies under six months should then be diluted with boiled water by 25 per cent (¼ water, ¾ milk) then 3.5g of sugar added to each 100ml of diluted milk, to restore the energy content. Advice from a mother with practical experience is that for vitamins A, C and D, one level teaspoonful a day of orange syrup and cod liver oil should be added. It is also necessary to supplement with folic acid, in which goats' milk is particularly deficient. A folic acid suspension may be added, 2.5ml twice a day in the bottle.

When solid food is added to the diet, all forms of cows' milk must be scrupulously avoided, which means reading the ingredients labels of all purchased foods. Unless goats' milk and dairy products can be used instead, there is a danger of shortage of some of the nutrients found in milk, such as calcium. Help can be obtained by allergy sufferers or their families, faced with this daunting task. Several useful publications are listed in the Bibliography on p. 323.

The British Allergy Foundation, formed in 1991, is listed in Appendix 3 (p. 320).

APPENDIX ONE

Recorded Milk-yields of the Dairy Breeds

It is difficult to obtain meaningful milk-yield figures, as the British Goat Society have published only yields reaching 1,000kg in a maximum of 365 days (2,000lb prior to metrication). It is believed that only about a quarter of recorded goats reach this figure and there are many which are not recorded. The Society has plans to publish all recorded yields in the foreseeable future.

However, for information, the maximum recorded yields for each breed and grading up goats, for seven of the years between 1935 and 1990, are given in Table 21. The last column gives average recorded yields for each breed for 1983. Yields have been converted from lb into kg where necessary.

Table 21 Maximum recorded milk yields 1935 to 1990

Breed	*1935*	*1945*	*1953*	*1965*	*1975*	*1985*	*1990*	*Average '83*
British Saanen	2,419	2,263*	1,916	2,453	2,156	2,534	2,617	1,261*
Saanen	1,691	2,352	1,773	1,542	1,879	1,829	1,697	1,153*
British Toggenburg	1,692*	1,583*	1,963	1,553	1,458	2,433	2,131	1,180*
Toggenburg	1,148*	937*	1,139	1,046	1,312	1,655	1,948	1,047*
British Alpine	1,648	1,822*	1,749	2,047	1,638	1,965	1,891	1,099*
Anglo-Nubian	1,348*	1,018	1,238	1,458	1,627	1,927	1,910	1,040*
Golden Guernsey	—	—	—	—	—	1,155	1,443	992*
Herd Book (British)	2,377*	2,430	2,846*	2,189	2,267	2,369	2,965	1,235*
Foundation Book	1,329*	1,426*	1,496*	1,625	1,251*	2,063	1,925	—
Supplementary Register	1,519	1,590	—	—	1,262	1,864*	1,561	—
Identification Register	—	—	—	1,050*	1,690	1,916	2,188	—

*Not recorded for the full 365 days. The average figures are for goats who milked for a minimum of 270 days.

APPENDIX TWO

Addresses of Goat Organizations in the UK and Overseas

UNITED KINGDOM

The British Goat Society, 34–36 Fore Street, Bovey Tracey, Newton Abbot, Devon TQ13 9AD Tel. 0626–833168

The following organizations are affiliated to the British Goat Society, from whom a current contact name and address can be obtained:

Many regional, county and milk-recording clubs

Breed societies

Anglo-Nubian Breed Society
British Alpine Breed Society
British Boer Goat Society
British Cashmere Goat Society
British Saanen Breed Society
British Toggenburg Society
Golden Guernsey Goat Breed Society (on Guernsey)
Golden Guernsey Goat Society (on the mainland)
Saanen Breed Society
Toggenburg Breeders' Society

Special interest societies

Caprine Ovine Breeding Services (COBS)
Harness Goat Society

Other organizations

The Bagot Goat Study Group, Ramshill, Mockley, Tanworth in Arden, Warwicks B94 45BA Tel. 05644–2354

The British Angora Goat Society, Fourth Street, National Agricultural Centre, Stoneleigh Park, Kenilworth, Warwicks CV8 2LG Tel. 0203–696722, Fax. 0203–696729
There are 13 regional Angora groups; contact names may be obtained from BAGS

The British Cashgora Association, Highfield, East Torrington, Lincoln LN3 5SD Tel. 0673–858730

The Coloured Angora Goat Association, Horton Park Farm, Horton Lane, Epsom, Surrey KT19 8PT Tel. 0372–743984

The English Goat Breeders' Association, Churchgate House, George Nympton, South Molton, Devon EX36 4JE Tel. 076957–2660

The Goat Veterinary Society, Milby, Brown's Spring, Potters End, Berkhamsted, Herts HP4 2SQ Tel. 0442–863101 (day), 0442–871845 (evening)

The Pygmy Goat Club, Fry's Cottage, Rosecare, St Gennys, Bude, Cornwall EX23 0BE Tel. 0840–3773

The Rare Breeds Survival Trust, National Agricultural Centre, Kenilworth, Warwicks CV8 2LG Tel. 0203–696551 (registers Bagot goats)

The Scottish Cashmere Producers' Association Ltd, 8 Polton Bank Terrace, Lasswade, Midlothian EH18 1JL Tel./Fax. 031–654 1305

OVERSEAS

International Goat Association, 1015 S Louisiana St, Little Rock, AR 72202 USA Tel. 010–1–501–376–6836, Fax. 010–1–501–376–8906.

The IGA has contact names in many countries for which goat societies are not listed below.

America

American Dairy Goat Association, 216 Wachusett St, Rutland, MA 01543-2099 USA Tel. 010–1–508–886–2221, Fax. 010–1–508–886–6729

Australia

Dairy Goat Society of Australia, Federal Secretary, Kiama Stud Registration Centre, Box 189, PO Kiama, NSW 2533 Tel. 010–61–042–32–3333

Belgium

National Association of Goats, Milksheep and Dwarfgoat Breeders, Van Roy Straat 30, 2548 Lint, Tel. 010–32–3–4552676

Denmark

Danmarks Gedeavlerforening, v/ Ingrid Dam, Petersmindevej 15, DK – 4250 Fuglebjerg

France

Fédération Nationale des Eleveurs de Chèvres (FNEC)
Tel. 010–33–1–40045317

Association Nationale Interprofessionnelle Caprine (ANICAP)
Tel. 010–33–1–40045131

Institut Technique de l'Elevage Ovin et Caprin (ITOVIC)
Tel. 010–33–1–40045308

All at:

Maison Nationale des Eleveurs (MNE) 149 rue de Bercy, 75595 Paris Cedex 12

Institut Technique des Produits Laitiers Caprins (ITPLC), Ecole Nationale d'Industrie Laitière, 17700 Surgeres
Tel. 010–33–46072033

UPRA – Caprine, 15 avenue de Vendôme, 41006 Blois,
Tel. 010–33–54786317
(Upra stands for breed selection and promotion unit, i.e. Herd Book)

SOPEXA, 43 rue de Naples, 75008 Paris Tel. 010–33–1–42944100
(ADETEF [Diffusion Abroad of French Breeding Techniques] has been taken over by SOPEXA, the French equivalent of Food From Britain)

Germany

Deutscher Ziegenzuchterverband, Schwerstr. 21, D/7000, Stuttgart 70 Tel. 010–37–711–4592713

Arbeitsgemeinschaft Deutscher Ziegenzuchter e.V., Andreas-Hermes-Haus, Godesberger Allee 142–148, Postfach 200454, W-5300 Bonn 2 Tel. 010–49–228–81980

Greece

Pasages, Kifisias 16, 115 26 Athens Tel. 010–30–115 26

Ireland

Irish Goat Club and Irish Goat Producers' Association
Addresses from British Goat Society, to which both are affiliated.

Netherlands

Algemene Nederlandse Bond Van Geitenhouders, Eperweg 65, 8181 Ev Heerde Tel. 010–31–5782–2317

Nederlandse Organisatie Voor de Geitenfokkerij, Wekeromseweg 26, 6218 Se Ede Tel. 010–31–8380–17193

New Zealand

New Zealand Dairy Goat Breeders' Association, Secretary, Mrs J. Glover, Box 34, Hikurangi Tel. 010–64–0894329862

Norway

Sheep and Goat Breeders' Association, Park Veien 71, Oslo 2 Tel. 010–47–2444288

Portugal

Associacão Central Agricultura Portuguesa, Rua D. Dinis, 2, 1200 Lisboa Tel. 010–351–01–3882462/3881395

Spain

Asociación de Ramaders de Cabrum de Catalunya, Gran de Gracia 218,2° 1ª, 08012 Barcelona Tel. 010–34–3239–8811

Asociación Nacional Criadores de Caprino Raza Murciana-Granadina, Caserio de San Pedro, 18220 Albolote, Granada

Asociación Nacional Criadores de Ganado Caprino Raza Segurena, Campanas 2, 18830 Huescar, Granada

Asociación Espanola de Criadores de la Cabra Malagueña, Alas de Lezo 2, 1°E, 29011 Málaga

Sweden

Swedish Association for Livestock Breeding and Production, Hållsta, 63184 Eskilstuna Tel. 010–46–16–163400

Switzerland

Committee of Swiss Cattle Breeders' Associations, Villettemattstrasse 9, CH–3000, Bern 14 Fax. 010–41–31–260880

Schw. Ziegenzuchtverband, Moosfang, CH–3783, Grund

Turkey

Tif Tik Birlik, Strazburg Cad 3/1–2–3, Sihhiye, Ankara

Note: There are organizations for breeders of fleece-bearing, as well as dairy, goats in Australasia but due to difficult times in the industry, permanent addresses could not be found

APPENDIX THREE

British Suppliers of Requisites for Goatkeepers and Useful Addresses

Note: It is not possible to include all the suppliers and useful addresses. The publishers would be interested to receive suggestions for additions.

Suppliers

All equipment

Lincolnshire Smallholders Supplies Ltd, Willow Farm, Thorpe Fendyke, Wainfleet, Lincs PE24 4QE

All small equipment: milking pails, disbudding irons, tethers, collars, etc.

Fred Ritson, Goat Appliance Works (est. 1938), Longtown, Carlisle CA6 5LA

Smallholding Supplies, Priory Road, Wells, Somerset BA5 1SY

A wide range including dairying equipment and vitamins

Goat Nutrition Ltd, Unit C, Smarden Business Estate, Monks Hill, Smarden, Ashford, Kent TN27 8QJ

Electric fencing

Flexinet, Unit C, Chancel Close, Trading Estate, Estate Avenue, Glos G44 7SH

Farm business advice

Genus, Westmere Drive, Crewe, Cheshire CW1 1ZD Tel. 0270–536536

Fencing

Drivall Ltd, Narrow Lane, Halesowen, West Midlands B62 6PA

Garlic, herbs, etc.

Dorwest Herb Growers, Freepost, Bridport, Dorset

Lucerne and other feeds

Dengie Crops Ltd, Hall Road, Southminster, Essex CM0 7JF

Minerals

Tithebarn Products Ltd, Southport, Lancs

'Optimalin' for tanning skins
Watkins and Doncaster the Naturalists, Four Throws Cottages, Conghurst Lane, Hawkhurst, Kent TN18 5DZ

Pots and lids
Cockx Sudbury Ltd, Unit 9, Alexandra Road, Sudbury, Suffolk CO10 6XH

Rennet, colouring, etc
Chr. Hansen's Laboratories Ltd, Reading RG2 0QL

Rugs for goats
Gillrugs, Old Cider Lodge, Kilmington, Axminster, Devon

Seeds
S. M. McArd, 39 West Road, Pointon, Sleaford, Lincs NG34 0NA

Useful addresses

Agricultural Development and Advisory Service (ADAS), Great Westminster House, Horseferry Road, London SW1P 2AE
Agricultural Training Board (ATB), Summit House, Glebe Way, West Wickham, Kent BR4 0RF
The British Allergy Foundation, St Bartholomew's Hospital, West Smithfield, London EC1A 7BE Tel. 071-600 6127
External Relations and Trade Promotion Division, MAFF, Room 313, 10 Whitehall Place, East Block, London SW1A 2HH (Also for other MAFF Divisions)
Farm & Rural Buildings Centre Ltd, National Agricultural Centre, Stoneleigh, Kenilworth, Warwicks CV8 2LG (Also for Royal Agricultural Society of England (RASE))
Goat Advisory Bureau, Water Farm Goat Centre, Stogursey, Bridgwater, Somerset
National Farmers Union, Agriculture House, 25–31 Knightsbridge, London SW1X 7NJ
United Kingdom Dairy Association, Giggs Hill Green, Thames Ditton, Surrey KT7 0EL (Also for Milk Marketing Board and Specialist Cheesemakers' Association)
Veterinary Medicines Directorate, Woodham Lane, New Haw, Weybridge, Surrey KT15 3NB

Bibliography

General

Devendra, C. and Burns, Marca, *Goat Production in the Tropics*, 2nd edn (Commonwealth Agricultural Bureau, Slough, 1983)

Gall, C. (ed.), *Goat Production* (Academic Press, 1981)

Halliday, J. and J., *Practical Goat-keeping* (Ward Lock, 1982)

Hetherington, L. U., *Home Goat Keeping* (E.P. Publishing, 1981)

Hetherington, L. U., *All About Goats*, 3rd edn (Farming Press Books, 1992)

International Goat Association, Journal of, Small Ruminant Research (Elsevier, Netherlands, 1988 onwards)

International Development Research Centre, *Small Ruminants Research & Development in the Near East* (IDRC–MR237e, 1990)

Islay & Jura Goat Society, *Goat Husbandry Survey Report* (Islay & Jura Goat Society, 1985)

Mowlem, A., *Goat Farming*, 2nd edn (Farming Press Books, 1992)

Neal, Jenny, *An Introduction to Goatkeeping* (Charter Magazines Ltd, 1988)

Neal, Jenny, *Goatkeeping for Profit* (David & Charles, Devon, 1988)

Proceedings of the IV International Conference on Goats, Brasilia, 1987 (International Goat Association)

Ross, Patricia, *Goats, a Guide to Management* (The Crowood Press, Wilts, 1989)

RSPCA, *Goats* (RSPCA, Horsham, 1987)

Thear, Katie, *Goats & Goatkeeping* (Merehurst Press, London, 1988)

UFAW (United Federation for Animal Welfare), *Management & Welfare of Farm Animals, the UFAW Handbook*, 3rd edn (Baillière Tindall, 1988)

Wilkinson, J. M. and Stark, Barbara, *Commercial Goat Production* (BSP Professional Books, 1987)

Breeds

Breeds of Goats (British Goat Society, 1990)

Handbooks of Breed Societies affiliated to the British Goat Society

Mason, I. L., *World Dictionary of Livestock Breeds*, 3rd edn (CAB International, 1988)

Feeding and Nutrition

Goodwin-Jones, R., *An Introduction to Trace Elements in Pastures* (Trace Element Services, Carmarthen, Dyfed, 1988)
Hunter's Guide to Grassland, Hunter's Guide to Grasses, Clovers and Weeds (Hunter's of Chester)
MAFF, *Poisonous Plants and Fungi* (HMSO Books, Norwich, 1988)
MAFF, *UK Tables of Feed Composition and Nutritive Value for Ruminants*, 2nd edn (Chalcombe Publications, Kent, 1992)
Morand-Fehr, P. (ed.), *Goat Nutrition* (Pudoc Wageningen, 1991, UK distrib. CAB, Slough)
Orskov, B., *The Feeding of Ruminants, Principles and Practice* (Chalcombe Publications, Kent, 1987)
Sheldrick, R. et al., *Legumes for Milk and Meat* (Chalcombe Publications, Kent, 1987)

Fencing

Agate, Elizabeth, *Fencing* (British Trust for Conservative Volunteers, 1986)

Fleeces

Morris, J. et al., *Goats for Fibre* (National Angora Stud, Cornwall PL30 4DW, 1987)
Russel, A. (ed.), *Scottish Cashmere* (The Scottish Cashmere Producers' Assn, Midlothian, 1987)
Russel, A. (ed.), *Scottish Cashmere, the Viable Alternative* (The Scottish Cashmere Producers Assn, Midlothian, 1990)
Ryder, M., *Cashmere, Mohair and other luxury animal fibres for the Breeder and Spinner* (White Rose II, Southampton, 1987)
Webb, Jacky and Saunders, Betty, *Mohair from Goat to Garment* (The Mohair Centre, Brickfield Farm, Laughton Road, Chiddingly, Lewes BN8 6JG, 1992)

Health and Disease

Bell, C. et al., *The Zoonoses, infections transmitted from animals to man* (Edward Arnold, 1988)
Dunn, P., *The Goatkeepers Veterinary Book* (Farming Press Books, 1982)
Journal of the Goat Veterinary Society

Macleod, G., *Goats: Homoeopathic Remedies* (C. W. Daniel Co Ltd, Essex CB10 1JP, 1991)

Matthews, J. G., *Outline of Clinical Diagnosis in the Goat* (Wright, 1991, distrib. Marston Book Services Ltd, Oxford)

Mowlem, A., *The Goat Market for Animal Health Products* (PJB Publications Ltd, Surrey, 1989)

Meat

Devendra, C. (ed), *Goat Meat Production in Asia* (IDRC–268e, 1988)

Grainger, Mary, *The Marketing of Goat's Meat* (NFU Marketing, Lincs, PE9 2DN, 1986)

Milking, Milk and Dairy Produce

Alfa-Laval, *Machine Milking of Dairy Goats* (Alfa-Laval, Cwmbran, Gwent, 1984)

Ash, Rita, *Cheesecraft* (Tabb House, Padstow, Cornwall, 1983)

Bennett, Carol, *Asthma and Eczema. A Special Diet Cook Book* (Thorsons, 1989)

Black, Maggie, *Home-made Butter, Cheese and Yoghourt* (EP Publishing Ltd, 1977)

Burton, H. (editorial board), *Short-Shelf-Life Products* (Society of Dairy Technology, Cambs, 1986)

International Dairy Federation, *Production and Utilisation of Ewes' and Goats' Milk*, IDF Bulletin No. 202 (UK distrib: United Kingdom Dairy Association, Thames Ditton, 1986)

Le Jaouen, J-C, *The Fabrication of Farmstead Goat Cheese* (*Cheesemakers' Journal*, PO Box 85, Ashfield, MA 01330, USA, in English 1987)

Locke, D. and T., *Cows' Milk Intolerance, an Alternative* (Great Glen Foods, PH33 6RZ, Scotland, 1991)

MAFF Code of Practice, *The Hygienic Production of Goats' milk*, BL5677 (MAFF pubns, London SE99 7TP, 1988)

Mowlem, A., *The UK Dairy Goat Industry* (Food From Britain, London SW8 5NQ, 1990)

Richmond, Sir Mark (chairman), *The Microbiological Safety of Food*, I, II (HMSO, London, 1990/1)

Miscellaneous

COGNOSAG, Workshop reports of Committee on Genetic Nomenclature of Sheep and Goats (Lavoisier University, Paris, 1986/7)

Davis, S., *The Archaeology of Animals* (Batsford, London, 1987)

Evans, G. and Maxwell, W. M. C., *Salamon's Artificial Insemination of Sheep and Goats* (Butterworth, 1987)

King, J. W. B. (ed), *Directory of Current Research on Sheep and Goats* (CAB International, 1988)

Thear, Katie, *Curing Skins* (Broad Leys Publishing, Essex CB11 3SP, 1981)

Whitehead, G. K., *The Wild Goats of Great Britain and Ireland* (David & Charles, Devon, 1972)

Also the publications of the British Goat Society, British Angora Goat Society, Scottish Cashmere Producers' Association, and other organizations listed in Appendix 2

Official publications, concerned with new legislation, etc.

MAFF Food Safety Directorate Information Bulletin (monthly). Free from: The Editor, Food Safety Directorate Information, MAFF, Room 303a, Ergon House, c/o Nobel House, 17 Smith Square, London SW1P 3JR Tel. 071–238 6550

The following are obtainable, free, from: MAFF Publications, London SE99 7TP Tel. 081–694 8862

Code of Good Agricultural Practice for the Protection of Air (PB 0618)

Code of Good Agricultural Practice for the Protection of Water (PB 0587)

Also similar Code for Soil (in preparation for 1993 publication)

Code of Recommendations for the Welfare of Goats (PB 0081), (if you look after goats you have, by law, to have a copy of this Code), also welfare leaflets PB0450 and PB0241

Code of Practice, The Welfare of Animals in Livestock Markets (PB 0409)

Various reports of the Farm Animal Welfare Council (e.g. PB 0124, PB 0687, £2.95)

A Guide to Clean Milk Production (cows) (PB 0341)

Code of Practice, The Hygienic Production of Goats' Milk (BL 5677)

1992 and You, Advice for the Meat Trade (leaflet PB0653)

Red Meat (PB 0654)

Poultry Meat (PB 0655)

Game Meat (PB 0656)

(Information about the Single Market Rules)

The Food Safety Act 1990 and You, a guide for the Food Industry (PB 0351)
A guide for caterers and their employees (PB 0370)
A guide for farmers and growers (PB 0371)
Best Before and Use By, a guide to the changes (PB 0411)
Numerous Food Safety booklets are also available

From Health Publications Unit, No. 2 Site, Heywood Stores, Manchester Road, Heywood, Lancs OL10 2PZ (both free of charge):
Stay Safe (ref FHI/E), brief explanation of new food temperature controls for cheese, etc.
Practical Food Safety for Businesses (HACCP – Hazard Analysis Critical Control Point)

From MAFF Divisional Veterinary Offices
Sheep and Goat Health Scheme Prospectus and explanatory leaflets
Copies of the legislation behind all these leaflets is available (variously priced) from HMSO, London, or regional branches. London enquiries: 071–873 0011, orders: 071–873 9090.

Index